Climate Diplomacy from Rio to Paris

Climate Diplomacy from Rio to Paris

The Effort to Contain Global Warming

WILLIAM SWEET

Yale UNIVERSITY PRESS

New Haven and London

Yale University Press books may be purchased in quantity for
educational, business, or promotional use. For information, please
e-mail sales.press@yale.edu (US office) or sales@yaleup.co.uk
(UK office).

Set in Minion type by IDS Infotech Ltd.
Printed in the United States of America.

Library of Congress Control Number: 2016948021
ISBN 978-0-300-20963-1 (paperbound)

A catalogue record for this book is available from the British Library.

This paper meets the requirements of ANSI/NISO Z39.48–1992
(Permanence of Paper).

10 9 8 7 6 5 4 3 2 1

To
Anna and Jacob, Victoria and Luke . . .
and their whole damned generation

Contents

Preface ix

Part I: The Stakes

ONE. Can Catastrophic Climate Change
Be Averted? 3
TWO. What Else Is at Stake? 18
THREE. Can Diplomacy Deliver? 37

Part II: The Players

FOUR. The Superpowers 61
FIVE. BRICs, BASICs, and Beyond 80
SIX. Sentimental Attachments, Existential Threats 98

Part III: The Action

SEVEN. The Road to Rio 119
EIGHT. Rio and Kyoto 126

NINE. Copenhagen 144

TEN. The Road to Paris 162

Epilogue: The Paris Agreement 170

Appendixes 185

Notes 191

Selected Bibliography 219

Acknowledgments 227

Index 231

Preface

At the beginning of April 2015, two Dutch polar researchers and explorers, Marc Cornelissen and Philip de Roo, set off on cross-country skis to investigate an area of thinning ice in northern Canada known as the Last Ice Area. Toward the end of the day on April 27, Cornelissen left a cheerful voicemail back home saying that he and his companion were having to ski in their underwear because of unexpectedly warm conditions. He also said that they might have to take a detour to get to their ultimate destination, Bathurst Island, because of unexpectedly thin ice. That would be the last such message. The next day the Royal Canadian Mounted Police received an emergency message from the two-man team, and when a pilot surveyed the area, he spotted the pair's sled dog but not the explorers. One corpse subsequently was recovered.

That same month, coincidentally, *Harper's Magazine* published an article, "Rotting Ice," by the intrepid nature writer and naturalist Gretel Ehrlich, describing repeated visits she had paid to seal-hunting Inuit in Greenland's far north, people living in some of the world's most remote villages. Because of ever widening open waters and ever scarcer ice floes capable of supporting seals, the native men and women were finding it impossible to go on with their traditional lives, in which the center from time immemorial had been the annual seal hunt. Their children, instead of being initiated into the fine arts of the hunt, were having to head south to get vocational educations to become auto mechanics, electricians, or plumbers. Surveying the big picture, a veteran of Ohio State University's Byrd Polar and Climate Research Center told Ehrlich: "The ice sheet is melting at an accelerated pace. It's not just surface melt but the deformation of the inner ice. The fabric of the ice sheet is coming apart because of increasing meltwater infiltration. Two to three hundred billion tons of ice are being lost each year. The last time atmospheric CO_2 was this high, the [global average] sea level was seventy feet higher."

Wherever on earth the effects of global warming are at their most acute—whether it is in the high-latitude regions of the Arctic and Antarctic or the topographically highest regions of the Andes and Himalayas—it is the same story: The ill effects of climate change are becoming dramatically worse dramatically faster than even the leading experts have expected. Scenarios that were considered almost outlandish just years ago, more fit for science fiction than for serious scientific consideration, now are matters of active concern. The total collapse of Antarctica's Ross Ice Shelf, an area the size of France, or the complete hiving off of Greenland's ice cover—these are developments, were they to occur abruptly, that could render virtually every major coastal city of the world uninhabitable from one decade to the next.

As scientists have started to think the unthinkable, there has been a subtle but distinct shift in attitude among members of the general educated public as well. Not so long ago, in the face of any uncommonly extreme weather event, it was the almost universal common wisdom to say that although the event might be consistent with global warming predictions, it of course could not be blamed squarely on global warming. Now, faced with such events, the common wisdom increasingly is to attribute them to climate change unless the contrary can be scientifically proven and even if expert opinion is expressly suggesting the opposite. Scientists may say that the drought that's been ravaging California could be just a random hundred-year event, but politicians, big media, and the general public appear not to care much. And that seems to be because most of us are coming to feel, whether or not the drought is a direct result of global warming, that it might as well be—we feel that it is sending a message that we ignore at our peril.

With that different sense of threat has come, too, an awareness that we in any one country cannot head off climate catastrophe all by ourselves, a realization that averting a cataclysm will require the combined and coordinated efforts of the whole world. Of course, that view is not universally held among Americans or generally, but it has come to be quite distinctly the attitude among people who actively worry about climate change. This is why hundreds of thousands took to the streets of New York City in September 2014, when the UN secretary-general convened a one-day summit to galvanize support for a strong international climate agreement. The specific purpose of the summit had been to gather world leaders, get them to focus on the climate problem, and inspire them to stronger collective action. But the public demonstration that was organized independently of the United Nations ended up outclassing the official event. The secretary-general himself joined the crowd—estimated at about a quarter of a million—embracing its favored slogan of the day, "There is no planet B."

A casual observer of the demonstration might have supposed that it really was aimed specifically at the United States and US policymakers, not the global community of climate diplomats. But that was not the case. In an informal but rather carefully randomized written survey I took during the demonstration, almost all respondents agreed that they wanted to bring pressure on the nations of the world to address climate change more aggressively. Though all of them said that the United States should take stronger action on climate change regardless of what anybody else did, 94 percent of them considered coordinated international action essential. What was more, high fractions of the demonstrators proved to be quite well informed about which leading countries had been playing a constructive role in climate talks and which had been more obstructive than helpful. Two thirds of them had formed specific opinions about the positions the United States had been taking in global climate negotiations (see appendix 1).

This short book is addressed to all students of climate policy, whether informal or formal, and proceeds from the premise that an informed public is the best single guarantor of sound public policy. The book's thesis is that climate negotiations, contrary to an opinion very widely held at present, can work and have worked. The shortcomings in the outcomes of negotiations are for the most part not the fault of the negotiating process, as such, but of the major participants in the process, who often have not played the diplomatic game to best possible effect.

The focus of the book is strictly on the effort to arrive at universally agreed-upon rules about how all the states of the world are to address global warming. Its purpose is to provide a short analytic history of global climate negotiations, from their beginnings in the early 1990s to the present day. It is an exercise in contemporary history, not a work of general social science or a study of what goes by the name of "global climate governance," a much broader topic. In describing the mechanics of how global climate policy has come to be formulated, it puts equal emphasis on fundamental social-economic forces and on the vagaries of chance event and exceptional personality—the essentially unpredictable elements that can make important things happen that otherwise might not have happened.

This being a book about ongoing developments, its method and philosophy are unabashedly journalistic. The book relies as much on direct observation and on conversations with knowledgeable people as on written documents, and it treats those interviewed not just as sources but as subjects of the story as well. Whether individuals are negotiators, scientists, policy intellectuals, or citizen activists, their actions and reactions make up the political chemistry of climate diplomacy.

As a historian and journalist writing about climate diplomacy, I take for granted that many important issues can be addressed only provisionally and sometimes speculatively, pending declassification of documents and publication of candid memoirs or letters. That is to say—and I apologize in advance for resorting to a cliché—the book is but that proverbial "first draft of history," written for those who think that something of the sort might be helpful.

Like journalists generally, I try to take no sides. My only allegiance is to what the historian Peter Gay called the party of humanity, the community of those who believe knowledge can be brought to bear to improve the lot of humankind. My central concern here is how well or poorly that cause is being advanced in diplomatic negotiations—in how well or poorly the game is being played.

Though high history this most assuredly is not, it does presume that climate diplomacy deserves to be treated with the same respect that we traditionally accord matters of war and peace. We members of the general public may not be thinking today of climate negotiations the same way we have thought of the great triumphs and failures of diplomatic history in the past—the Congress of Vienna in 1815, say, or the Cold War's end in 1989, the Versailles Treaty or the Oslo process—but we should. Whether catastrophic climate change is headed off will depend greatly on what happens in diplomatic negotiations. If those negotiations fail, it will be only a matter of time until states are at each other's throats, as the circumstances we have depended on crumble all around.

Climate Diplomacy
from Rio to Paris

I

The Stakes

All students of diplomacy will agree that diplomatists have often progressed further than politicians in their conception of international conduct, and that the servant has more than once exercised a determinant and beneficial influence upon his master.

—HAROLD NICOLSON, *Diplomacy* (1939)

1

Can Catastrophic Climate Change Be Averted?

The stated purpose of the ground-setting United Nations Framework Convention on Climate Change, adopted at Rio de Janeiro, Brazil, in 1992, was to prevent dangerous human-made climate change. So in due course parties to that treaty, in which membership is virtually universal, would have to specify what was meant by dangerous climate change. That process reached its fulfillment only seventeen years later, at the climate conference in Copenhagen, Denmark, where the concluding accord acknowledged a scientific consensus that global warming in the twenty-first century should be limited to no more than two degrees Celsius (2°C). The Copenhagen conference was a major disappointment to everybody who had been hoping for and expecting conclusion of a strong new climate agreement. But its recognition of the 2°C threshold was one of its major saving graces, noted Richard Kinley, the number-two person at the Framework Convention's secretariat in Bonn, Germany.[1]

Germany, as it happened, had played a key role in the diplomatic process that led to the 2°C consensus at Copenhagen. First, Chancellor Angela Merkel persuaded the European Union to adopt the 2°C goal as part of the 20/20/20 program it agreed upon in 2007, calling for a 20 percent reduction in greenhouse gas emissions from 1990, 20 percent renewable energy, and a 20 percent improvement in energy efficiency—all by 2020. Two years later, Merkel was again instrumental in getting the United States to sign on to the 2°C target, this time at a Group of Eight (G8) meeting of the leading industrial countries.[2]

Nobody has ever pretended that the 2°C goal is absolutely scientific or apolitical; obviously it is the outcome of a bargaining process. Yet it has a persuasive intuitive logic, which is no doubt why it won the day. For one thing, 2°C represents the maximum amount, during the past million years, that the average global temperature has exceeded temperatures prevailing at the beginning of the twentieth century. German negotiators took the position that warming, to be on the safe side, should be kept within the historical bounds experienced in the last thousand millennia.[3] Indeed, computer models suggested that beyond 2°C there might be some disconcerting effects. For example, the entire Greenland ice sheet could begin an unstoppable melt, raising the world's sea levels an estimated 23 feet. "Risking a loss of the whole Greenland ice sheet was considered a no-go area," Stefan Rahmstorf of Germany's Potsdam Institute for Climate Impact Research told Justin Gillis of the *New York Times*.[4]

At the time German negotiators were advancing that position, a scientific consensus had developed that 2°C would correspond to an atmospheric concentration of carbon dioxide (CO_2) of roughly 450 parts per million (ppm). At a conference held at the Hadley Centre for Climate Prediction and Research, in Exeter, England, scientists agreed that 2°C of warming would be about as likely as not at 450 ppm and almost a dead certainty at 550 ppm; to be absolutely confident of stabilizing temperatures at 2°C one would have to prevent atmosphere concentrations of CO_2 from exceeding 400 ppm. Previously, scientists had thought 2°C warming would be overwhelmingly likely only at 550 ppm, where concentrations would be roughly twice what they were before the Industrial Revolution began in the mid-eighteenth century.[5]

Because of its simple numerical relationships with atmospheric greenhouse gas concentrations and with preindustrial and prehistoric temperature variations, 2°C was a rather easy notion to get one's head around, though it may have struck some people as overly cautious (and others as not ambitious enough). On the face of it, 2°C of warming—3.6 degrees Fahrenheit—does not sound like a very big change. But that is partly because so many of us unconsciously tend to think in terms of our usual temperatures: In temperate zones warming of 2°C might be interpreted as the difference, say, between 70°F and 73–74°F. But our everyday experience is the wrong frame of reference here. The average preindustrial temperature of the Earth was just under 14°C, and so warming of 2°C represents an increase from 14°C to 16°C, or from 57°F to almost 61°F. The world has warmed roughly 0.8°C in recent decades; so a limit of 2°C implies that warming in the remainder of this century should be no more than 150 percent greater than the warming that has occurred so far.

Thus, contrary to appearances, 2°C of warming is not a small change, and by the same token, it will not be easy to keep the world within that limit. Beginning in 2015, the Earth's average atmospheric concentration of carbon dioxide crossed the 400 ppm threshold, a level unprecedented in the last three million years. The level is increasing by about 2 ppm per year, and at that rate, it will take the world just twenty-five years to reach the 450 ppm limit. To keep the level below that limit and warming at 2°C or less, the total quantities of greenhouse gases being pumped into the atmosphere yearly need to start coming down much sooner than a quarter century from now—a huge challenge, by any reasonable estimate.

In September 2013, four years after the 2°C limit was recognized at Copenhagen, the Intergovernmental Panel on Climate Change (IPCC) published a calculation of how much carbon in total the world could emit if we were to stay below 2°C warming. The IPCC, a global organization of volunteer scientists, set the cap at one trillion metric tons of carbon and found that since the Industrial Revolution began, we have already emitted more than half the amount we ever will be entitled to spew into the atmosphere. That would imply that one third of the world's oil reserves, half its natural gas deposits, and 80 percent of its known coal would have to go untapped.[6]

Every fundamental principle established in twenty-five years of climate diplomacy has immediately come to be contested, to be subjected to a long process of ongoing interpretation and review. And why should that be otherwise? The fundamental principles of the US Constitution, and every other major document of its kind, have been continuously disputed, reviewed, and reinterpreted—and always will be. So it is hardly surprising that even before the 2°C limit was "acknowledged," in the curious language of the 2009 Copenhagen Accord, it was being critiqued in the press, in scholarly literature, and—not least—in the streets. Inspired and to an extent organized by environmental writer and activist Bill McKibben and the global organization he created, 350.org, demonstrators and nongovernmental lobbyists took the position at Copenhagen that 450 ppm and 2°C were inadequate criteria of dangerous warming.[7] What we needed to shoot for instead, they argued, was to work ourselves back to 350 ppm, implying immediate, sharp, and ubiquitous cuts in greenhouse gas emissions.

A contrary line of thought, and one that gathered considerable force in the five years following Copenhagen, held that 2°C/450 ppm was simply unattainable and that stubborn insistence on it was taking us into a dangerous dream world. That point of view, though widely shared in expert circles, came to be associated particularly with David G. Victor, a professor of international relations at the University of California, San Diego. In October 2014, Victor and Charles F. Kennel, a former director of the Scripps Institution of Oceanography in San Diego, published an article in *Nature* magazine in which they declared the 2°C goal to be "effectively unachievable."[8] They argued that taking one single measure such as temperature as a measure of climate change was simplistic and that the world needed instead to formulate a menu of actions needed, like the twenty-one targets and sixty detailed indicators the United Nations spelled out under the rubric of its Millennium Development Goals. Focusing narrowly on temperature, despite the absence of a "lockstep relationship" between climate change and warmth, only "allowed politicians [and "some governments"] to pretend they are organizing for action when, in fact, most have done little," they said—twice, actually, in almost exactly the same words, evidently for emphasis.

Both publicly and privately, many leading experts on climate policy agree with Victor and Kennel that the 2°C goal is unachievable, though there is far from universal agreement as to whether the big stumbling block is technological, economic, or political. Yvo de Boer, chairman of the UN organization that held the Copenhagen conference, said in 2013, looking ahead to the Paris climate meeting scheduled for December 2015, that nothing consistent with less than 2°C warming could possibly be negotiated: "The only way that a 2015 agreement [could] achieve a 2-degree goal [would be] to shut down the whole global economy."[9] Yet the contrast between that kind of view and opinions widely expressed and heard at the grassroots is stark. Among activists and nongovernmental organizers, it seems to be widely believed that the goal can be almost readily reached by a fast and determined transition from fossil fuels to a green economy. On this view, what is standing in our way is not the ambitiousness of the goal itself but rather, to put it crudely, the naked self-interest of the big oil, gasoline, and coal companies, the excessive influence of big business over government generally, and the very nature of capitalism itself.

When we see dramatic headlines saying, for example, that last week for the first time in history, for a short interval, Germany satisfied almost three quarters of its midday electricity demand with renewable fuels, or that last year nearly half the world's net investment in new energy capacity went for renewables, it is easy to imagine that we are just a short step away from a carbon-free world.[10] In the Scandinavian states and Finland, renewables now account for two-thirds of electricity generated on average; since 1995, the combined GDP of those states has grown 45 percent even as CO_2 emissions have dropped 17 percent.[11] Zero- and low-carbon sources of energy are indeed growing fast almost everywhere in the industrial world, and their potential is very far from exhausted. Opportunities for energy conservation and improved energy efficiency also are all around. But challenging limits and costs are apparent too.

Take wind. Starting in the 1980s and 1990s, with the introduction of new materials and design techniques borrowed from the aerospace industry and power control techniques exploiting large semiconductor devices, there was a technological revolution in windmills. The new wind turbines first were introduced widely in northern Germany and Denmark, and soon, with the added stimulus of policy, became ubiquitous. By the turn of the century just about every suitable site in Denmark's Jutland Peninsula and in the vast North German Plain had come to be occupied with turbine complexes. Now developers began to push offshore, with the construction of giant farms in the North Sea, around the coasts of the British Isles, and into the hilly and wooded areas of southern and southwest Germany. This is where the limits began to appear. Today, in scenic areas like the Hunsrück region, south of the Mosel River and west of the Rhine, the large turbine towers no longer seem such a stirring sight and sometimes border on being a blight. Driving at night, with the beacon lights of the high towers all around, one would think one was in a heavy industrial area, surrounded by a grid of transmission towers. During the day, hand-painted roadside signs say simply, "Enough" (*genug*). In England, where bird lovers often despise wind turbines, Friends of the Earth now spends much of its time trying to persuade citizens at the grassroots to accept construction of windfarms. Decades ago such organizing efforts focused on prevention of nuclear construction.

Certain technical limitations of wind have been much discussed and are well understood: Its intermittency, the difficulty of harnessing it to full effect without large-scale electrical storage capacity (as yet undeveloped), and its tendency to be most abundant far from where it is most needed. But if you ask why the United States has yet to build a single offshore wind farm and why proposed projects for Cape Cod and Long Island have been stalled for years, it comes down to aesthetic considerations—residents are torn between their impulse to go green and their desire to continue enjoying the blue skies and waters, unencumbered by unsightly turbine towers.

Developments in solar energy have been in some ways parallel, but with a lag. Here the underlying technological revolution had its roots in solid state physics and in the super-sophisticated manufacturing techniques developed in the semiconductor industry. After a slow and halting beginning, prices for photovoltaic cells came down very dramatically in the last decade, as Chinese manufacturers stormed the world market, taking advantage of generous European and North American solar production subsidies. The end result was that solar-generated electricity came to be cost competitive with coal and natural gas in optimal settings, matching wind's competitive breakthrough about fifteen or twenty years before.

Yet solar energy also suffers from technical limitations that are in most respects similar to those of wind, and there is another important limitation to boot: its hunger for land. A very positive aspect of wind energy, besides its economic competitiveness and almost completely green profile, is that turbine towers consume virtually no real estate. In Denmark, farmers plow their fields right up to the towers' bases. With solar energy, unfortunately, the situation is the reverse: Farms of solar cells cover a great deal of land, and when they are constructed in fertile agricultural areas, they quite obviously are occupying land that could be put to other profitable uses. In Iowa, for example, solar has yet to achieve cost parity with wind not because sunshine is scarce or wind is superabundant but because the state's fertile land is so enormously valuable.[12]

Close to big urban and industrial load centers, where electricity is most needed, suitable land for the large solar farms also is scarce and expensive. So, at least until further efficiencies in photovoltaic cells are achieved so that they can be more generally integrated into buildings' walls and roofs and their footprints on land are much smaller, their full potential will not be fully realized.

Such examples could be multiplied if we turned our attention to issues of energy efficiency and consumption, where near-term opportunities for carbon saving are greatest. Suffice it to say that application of current technology will almost always require retrofitting of installations and equipment, which is not cost-free. Nor is the ongoing process of innovation and research needed to sustain continuing improvements in all the ways energy is put to use.

In spring 2015, the Organisation for Economic Co-operation and Development's International Energy Agency (IEA), the world's premier authority on global energy trends, issued a report, "Tracking Clean Energy Progress 2015," surveying all the technologies critical to achieving the 2°C goal. Despite rather dramatic advances in such areas as wind and solar, it found that progress was falling short of what was needed in every single one of the goal-relevant fields. It singled out building heating as an area in which efforts were particularly inadequate and transportation as one in which trends were especially adverse. Reliance on inexpensive coal was found to be growing faster than ever, putting a squeeze not only on renewables but also on relatively low-carbon natural gas and virtually zero-carbon nuclear energy. In heavy industry—notably steel and cement production—current trends on both energy efficiency improvement and direct carbon emissions fell far short of what would be needed by 2025 to get on a 2°C track. The same went for trends in automotive fuel economy and the rates at which electric and hybrid-electric vehicles were being introduced. While the IEA found bright spots here and there—the opening in

October 2014 of the first commercial-scale coal-fired plant where carbon emissions would be captured, for example, and the largest number of nuclear reactors under construction in twenty-five years—generally its mood was gloomy. Pretty much across the board, it found that there was a need everywhere for much stronger policy pushes to achieve the penetration of low-carbon energy and the amount of energy savings needed to keep global warming below 2°C.

Can warming in fact be limited to 2°C? If it cannot, would it not be better to relax or reformulate the standard, so that we are saying we will do something we actually have a reasonable chance of accomplishing? After all, every major survey of the subject has found reasons for skepticism. The World Bank, in November 2014, determined that we already are "locked in" to 1.5°C warming and that if current trends were to continue, we would be on course for 4°C, "a frightening world of increased risks and global instability." The International Energy Agency's *Energy Technology Perspective 2014* said that 6°C warming is "where the world is now heading, with potentially devastating results." The UN Environment Programme (UNEP), in an elaborate study of the gaps between current emissions and what would be needed to stay under 2°C, found in 2014 that total world emissions would have to start heading downward very early in the next decade and be 15 percent lower by 2030 than in 2010—and 50 percent lower by 2050. In its 2015 update report, UNEP found that staying below 2°C would imply net-zero greenhouse gas emissions by 2060–75—any emissions would have to be counterbalanced by uptake of carbon dioxide by forests and agriculture, or the gases would have to be directly removed from the atmosphere by chemical means. Most telling of all, perhaps, was the 2014 study done under the auspices of the United Nations and led by Jeffrey Sachs, head of Columbia University's Earth Institute. Dubbed *Pathways to Deep Decarbonization,* the idea was to show how countries could adopt growth paths consistent with the 2ºC limit. But the report's effect may have been almost the opposite—to show in fact that the job was virtually undoable.[13]

Using a method the authors of the report called backcasting, they determined what emissions would have to be in the future to stay under the 2ºC limit: They derived an average per capita figure for global emissions and then used that figure as a "benchmark" for each country's suggested pathway. A target of 15 billion metric tons of carbon dioxide emissions in 2050 translated into a benchmark figure of about 1.6 metric tons carbon per person. Taken literally, that number would imply that the United States would have to cut its per capita emissions (17.6 metric tons per year) by a factor of ten by 2050. Germany and Japan (at about 9 metric tons) would have to reduce theirs by a factor of six; the United Kingdom by a factor of five, and France by a factor of four. China, having seen its per-person emissions surpass France's, would have to cut them by roughly the same amount that the advanced industrial countries of the European Union would be cutting theirs. Even relatively low-emitting large countries like Brazil and India—at 2.2 and 1.7 metric tons, respectively—would have to cut their emissions slightly to approximate the 1.6 metric ton target. All other developing countries would have to adopt growth strategies that somehow avoided taking them much above 1.6 metric tons.[14]

Thus, the *Pathways* report concluded, "Staying within [our] CO_2 budget requires very near-term peaking globally, and a sharp reduction in global CO_2 emissions thereafter." As things stand, "the world is on a trajectory to an increase in global mean temperature of 3.7ºC to 4.8ºC, compared to pre-industrial levels." Sachs himself has done nothing to conceal just how hard it will be to get off that 3.7–4.8°C path. In a major book about sustainable development published concurrently with the *Pathways* exercise, a chart vividly shows the tight relationship between per capita income and per capita energy consumption that has prevailed in modern industrial times. In effect, he wrote, "We must undertake a kind of 'heart transplant,' replacing the beating heart of fossil fuel energy with an alternative based on low carbon energy."[15]

It may be tempting to suppose the answer to that Herculean challenge is another Manhattan or moon shot project, geared to low carbon energy. (Sachs himself has made that suggestion.)[16] Without a doubt, investment in energy research and development, from smart grids to zero-net-energy homes, is much lower than it should be. But the crash programs to build an atomic bomb and get to the moon are not the best points of reference. Both involved all-out development of a well-defined handful of critical technologies, which in combination would almost certainly accomplish the desired mission.[17] In the case of the so-called green transition, the number of relevant technologies is enormously much larger—almost infinite, in the sense that we do not know what all the relevant technologies might turn out to be: We can only guess which will pan out in the foreseeable future, or what combination could keep us below 2°C warming.

The truth is, we do not know how we can possibly meet the 2°C limit. By the same token, however—and this point deserves equal emphasis—there is a reasonable chance that among the dozens of green technologies relevant to 2°C warming there will be game-changing breakthroughs in at least two or three of them in the next decades. A versatile and economically

viable way of storing electricity generated by intermittent renewable sources would instantly multiply the potential of wind
and solar; a radically better car battery or fuel cell could make
zero-emission automotive transport an instant reality; an economically viable way of producing cellulosic ethanol could
make automobiles carbon-neutral if not emissions free; a
smaller and inherently safer nuclear reactor would radically
improve prospects for low-carbon pathways in rapidly developing countries; a breakthrough in thin-film photovoltaics, where
dozens of innovative materials are being developed and evalu-
ated, would transform solar.

In July 2015, the world's largest steelmaker, ArcelorMittal,
announced that it was investing eighty-seven million euros at
a plant in Belgium to install a process that uses microbes found
in a rabbit's intestines to turn waste carbon monoxide into us-
able biofuel.[18] The new technology, invented by a small New
Zealand company, was a nice example of how something
wholly undreamed of can suddenly have a transformative
impact in some particular sector.

In considering our prospects of limiting warming to 2°C,
we also should not forget that history is unpredictable and that
economic projections almost never are realized. There are
examples in recent history of unexpected developments radically altering the energy picture. Before 2005, outside very
narrow expert circles, hardly anybody anticipated the dramatic impact of natural gas fracking on US energy supplies and
short-term greenhouse gas emissions (whatever one may think
of that technology). Victor was completely typical when he
wrote in 2004, "If we build more gas-fired power plants in
countries like the United States, where gas is already scarce,
where will we get more gas?"[19] Much the same goes for the sudden plunge in world oil prices in 2014–15, which caught even

leaders of the petroleum industry off-guard.[20] Whether such developments are rated net-positive or net-negative depends, in the case of natural gas, on how much weight is put on local environmental hazards and on near-term versus long-term greenhouse benefits, and, with oil, how much weight is put on the unfortunate impact of lower gasoline prices on development of zero-emission cars versus more benign effects on economic growth and coal demand.[21] Whether net-negative or net-positive, the point here is merely that recent gas and oil developments have been hugely important and quite surprising.

Things happen and conventional wisdom changes. Events that would be terrible misfortunes from every other point of view—economic collapse in a major emitting country, famine, plague—could be good luck in terms of keeping carbon emissions down. There could be changes in modeled estimates of how much warming is associated with any given level of emissions: It could be that we will have more headroom than scientists currently believe.[22]

It also is the case that some promising ways of making dents in emissions have not been fully exploited within the standard frameworks of climate diplomacy and policy. Eminent climatologists like V. Ramanathan at the Scripps Institution of Oceanography and James Hansen of NASA's Goddard Institute for Space Studies at Columbia University (now emeritus) have emphasized the desirability of attacking short-lived climate pollutants, which have been neglected, arguably, because there has been such a strong focus on the long-lived greenhouse gases, first and foremost carbon dioxide. Some of the short-lived climate pollutants like methane and ozone are among the six greenhouse gases specified in the Rio Framework Convention and therefore have been counted in national inventories of gas trends. But others, notably black carbon—basically soot from incomplete combustion of fossil fuels and biofuels—have not. Black carbon is a major health hazard and an important greenhouse gas. If its levels could be sharply reduced permanently, this would contribute significantly to achieving the 2°C goal.

Another significant source of greenhouse gases that has fallen between the cracks in global diplomatic efforts is international aviation and shipping. Somewhat inexplicably, they have not been accounted for to date in national inventories. Yet together they make up 2 or 3 percent of global emissions.

In the final analysis, the best single reason to stick with the 2°C limit is simply this: It is the strongest estimate scientists have been able to give us of the threshold where climate change will get really, really dangerous. If you are in a car hurtling toward a cliff or a brick wall and you realize that you may not be able to brake in time to save yourself or that abrupt braking may hurtle you off the road, that does not mean you do not slam on the brakes. You slam on the brakes and hope for the best. That in essence is what we need to do.

2

What Else Is at Stake?

In their critique of the 2°C warming standard, described in the previous chapter, David Victor and Charles Kennel did not actually say that it technically could not be met, only that it would not be met. But that represents a political judgment about diplomatic prospects, a matter where judgment is best left suspended. As for their view that the standard should be ditched in favor of a much larger family of climate change criteria, like the United Nations' Millennium Development Goals, would this not amount to "kicking the can down the road," as Joseph Romm immediately suggested?

Romm, a former US Department of Energy official, runs one of the major websites devoted to climate change and is widely considered one of the best-informed critics of climate policy. In a comment he posted on Climate Progress, he observed trenchantly that "adding more vital signs [as Victor and Kennel proposed] just gives people more things to argue about, so it is hardly a recipe for faster or more streamlined international action. Indeed, the whole Victor and Kennel approach would be an excuse for more dawdling."[1]

As Romm implied, the large number of fundamental divisive issues in climate policy is in itself a major obstacle to diplomatic progress. Introducing the UN Environmental Programme's *Emissions Gap Report 2014* and looking ahead to the next major round of negotiations, scheduled for December 2015, UNEP chief scientist Joseph Alcamo put it like this: "Although the time to the Paris conference is short, the list of issues to be decided is long."[2] And that is not to mention some important issues that would not be decided at Paris or even taken up, as well as some so fundamental that they will never be resolved once and for all.

The most basic of all quandaries is the conflict between economic development and greenhouse gas reduction, or "mitigation" in the jargon of climate diplomacy, which refers both to cutting carbon emissions with respect to some reference year and to cutting emissions relative to what they would be in the absence of deliberate policy action. (For simplicity, any kind of mitigation will be referred to henceforth as carbon cutting.)

Among leaders of the world's very poor and relatively poor countries, it is taken for granted that their number one priority is to raise their peoples' standard of living. Because of the historical lockstep links among economic growth, reliance on fossil fuels, and high greenhouse gas emissions, it also is often taken for granted that emissions cannot be cut in the near future without hampering development. Attempts are made by the major development assistance organizations like the World Bank and UNEP to soften the conflict—mitigate it, if you will—by arguing that traditional growth paths are not sustainable. And this is not just rhetoric. Such organizations make real efforts to identify sustainable growth paths and provide financial and technical assistance to help developing countries onto such paths. But such efforts, though completely laudable and sound in theory, are not entirely convincing from a political point of view. If you happen to be leading a country where the population is booming and millions are pouring into cities, desperate to find the kinds of jobs that industrialization and globalization promise, you are not going to focus much on the issue of whether this or that economic opportunity represents a path that will be sustainable in the long run. (In the long run, to paraphrase John Maynard Keynes slightly, the leader in question will be dead.)

Selwyn C. Hart is director of the UN Secretary-General's Climate Change Support Team. A native of Barbados with a background in economics and finance, he has served as the Caribbean's top climate negotiator and as the diplomatic leader of AOSIS, the Alliance of Small Island States. Hart does not minimize the gravity of the climate threat. Speaking in New York City at the New School on May 11, 2015, he warned that "we are the last generation that will be able to prevent the worst consequences of climate change" and complained that greenhouse gas emissions are growing faster than ever, outdistancing political efforts to curtail them. And yet when it comes to India, one of the countries whose emissions are expected to grow the fastest in the coming decades, Hart said that it would be grossly unfair to ask a nation with four hundred million severely impoverished people to limit its economic growth and emissions any time soon.

That captures the nub of the dilemma. The evidence is overwhelming that unless all major emitting nations join in emissions cuts during the next decades, the whole effort to head off catastrophe is doomed. William Nordhaus, a Yale economist who has written two major books about climate policy (among many other things), spelled this out compellingly in the more recent of the two books, *Climate Casino.* Nordhaus estimated that if all countries adopted efficient policies to cut greenhouse gas emissions, the aggregate long-term cost of meeting the Copenhagen 2°C target might be 1.5 percent of world income, which he called "modest."[3] But if only half the countries of the world participated in cuts, the long-term penalty in global income would double, to 3 percent. When account is taken of the fact that most countries would *not* adopt the most efficient possible carbon-cutting policies, the long-term aggregate costs would be even higher—perhaps twice as high. "If half the countries make no efforts to reduce emissions, substantial warming will be inevitable even if the other half of countries make maximum efforts," he concluded.[4]

Nordhaus went on to argue that it makes no sense to base climate targets on science alone, without taking costs into account, but here he trod on somewhat shakier ground. Of course, global targets cannot be set without any regard whatsoever to their economic viability. But as a practical matter, it is important to recognize that each country participating in negotiations will be primarily concerned about the particular trade-offs it faces between greenhouse gas reduction, on the one hand, and economic growth and jobs creation, on the other. In other words, any one country does not really care how big or small the total costs of addressing climate change are as long as its own costs are manageable and are outweighed by its own benefits. There will be more to say about this below. Suffice it to say here that Nordhaus was arguing from the fundamental premises of classical economics, which presume that the most efficient way is by definition the way to proceed, based on an accounting of long-term costs and benefits.

The German climate scientists Stefan Rahmstorf and Hans Joachim Schellnhuber, in a 2008 book, subjected that notion to a withering critique, under the heading "Is there an optimal climate change?"—meaning, is there an optimal amount of climate change that can be selected in terms of total costs and benefits?[5] Insuperable difficulties in that approach include, first, the impossibility of assigning monetary value to lost human lives, vanished species, and disappearing natural wonders. Second, there is the only approximate possibility of calculating all future climate damage and adaptation costs: Do we need to reckon the added probability of a Superstorm Sandy or Typhoon Haiyan not only in 2030 or 2050 but also in 2070 or 2090? Not least there is the philosophical issue of how much weight should be given the welfare of future generations, a question to which there is no one answer (and to which we shall also return).[6]

Arguably, the classical economist's approach to reconciling growth and effective climate policy is impractical and unsound. A contrary approach, most closely associated at present with the writer and public intellectual Naomi Klein, in effect wishes the conflict away. It holds that effective climate policy is flatly incompatible with the basic premises of neoliberal economics and indeed with capitalism itself.[7] But this makes everything both too hard and too easy: too hard because if heading off catastrophic climate change requires us first to get rid of capitalism, then there is no hope of avoiding disaster;[8] and too easy because it ignores that anticapitalist countries have embraced fast growth just as enthusiastically as their capitalist enemies, or even more so.[9] In both communist Russia and China, modernization and growth had equal place with social justice in the regimes' stated major aims.

It is true, to be sure, that globally coordinated climate policy is incompatible with the antiregulatory neoclassical economics that was dominant in the West from the 1980s to the first decade of this century. Klein is quite right that the fundamental suspicion toward climate policy on the part of business-oriented political conservatives arises mainly from an antagonism to any form of government planning. But even here there are exceptions. Margaret Thatcher, the intellectual and political godmother of neoliberalism, also happens to have been the first world leader to draw sharp attention to the urgent problem of global warming. Speaking to the Royal Society on September 27, 1988, Thatcher enumerated three major threats to the world's atmospheric chemistry to have emerged in recent years and ranked global warming the first and most important. The following year Thatcher would establish the Hadley Centre for Climate Prediction and Research at the United Kingdom's Meteorological Office, the world's oldest and most prestigious weather forecasting organization. Her anointed successor as prime minister would sign onto the Rio Framework Convention on Climate Change in 1992, and every British government since, whether Conservative or Labour, has been an unstinting world leader in carbon reduction efforts. Attitudes about neoliberal economic policy have had next to nothing to do with it.

The bottom line is simply this: Since Karl Marx, it has been generally recognized and accepted that a growth imperative is built into capitalism. But political followers of Marx, nonetheless, have almost always accepted the need for further fast growth—and sometimes it has been conservatives who have recognized some need for limits to growth. Some conflict between economic growth and the need for universal greenhouse gas reduction is fundamental and ultimately ineluctable. And so every time the representatives of the 195 countries that are party to the UN Framework Convention on Climate Change get together to discuss carbon cuts, they will be arguing about how to resolve that conflict, and whatever they agree on will be good only for some negotiated period of time. Long before the end of that time arrives they will all be arguing the same issue again, as there can be no final resolution of the tension.

Almost equally intractable is the tension between mandatory and voluntary approaches to cutting carbon. The 1992 Rio Framework Convention made a fundamental distinction between advanced industrial countries and the developing countries, and accordingly, the follow-on Kyoto Protocol of 1997 required only the advanced industrial countries to make cuts. Immediately, Kyoto aroused virtually unanimous antagonism in the American political classes, and during the next decade, while other industrial countries made cuts, the United States stayed on the sidelines. In US climate policy circles, Kyoto came to be almost universally dismissed as a failure, not mainly because the United States sat on its hands for fifteen years but because of China's escalating emissions and the Kyoto provisions exempting countries like China from making carbon cuts.

At the Copenhagen climate conference in December 2009, the United States expressly shut the door on the Kyoto process forever, obtaining instead an informal declaration that the world would now adopt an approach called "pledge and review"—an approach that Japan had advocated before Rio, with US support, and which had been rejected at Rio and Kyoto in favor of binding treaty cuts. Especially in US policy circles, the new approach soon came to be referred to as "bottom up" rather than "top down," not because of that distinction's intrinsic merit, arguably, but because this kind of rhetoric sounds good to the kinds of people who habitually oppose any strong international action on climate change.

Ideological opposition to action on climate change has come primarily from the policy think-tanks and university groups most firmly dedicated to the free-market, neoliberal approach to economic policy, as Klein has rightly observed.[10] People in this camp see any binding international action on climate as a Trojan horse for socialist central economic planning, whether in the old Soviet style or the contemporary Scandinavian mode. Because of that political chemistry, the language of "bottom up" voluntary cuts is enormously enticing to those seeking to win the support of climate skeptics. Even UN officials like Kinley and Hart use it. Still, in terms of diplomatic substance, the up-down distinction obfuscates as much as it clarifies.

Any international agreement with legal force is by nature top-down. Governments representing states commit to actions that they and their successors might otherwise not necessarily take. Whether commitments take the form of pledges or collectively agreed-upon cuts, unless they are in some sense binding, they are essentially meaningless. Admittedly, a pledge and review system may encourage governments to arrive at diplomatic positions in a more bottom-up way, by inviting all interested parties to get their two bits in, before positions are finalized. But no competent government takes a diplomatic position without sounding out key stakeholders or without taking their attitudes and interests into account. From this point of view, the pledging system merely encourages governments to formulate policy in a marginally more competent way.

To give credit where credit is due, in the six years from Copenhagen to Paris, the pledging system did seem to move the world toward universal participation in carbon cuts. It did so by nudging countries along, to use a term popularized in policy circles by the University of Chicago scholar Cass Sunstein. But that process will be a proven success only if in the end the collective pledges are sufficient to get the world onto a 2°C path, and only if the end result is seen to have some real legal and moral standing. Hard bargaining conducted in the name of firmly shared goals, not just nudging, will be required.[11] If the pledge-and-review system inaugurated at Copenhagen fails to produce an agreement with adequate commitments, or if the agreement simply lacks teeth, then questions naturally will be raised as to whether there might have been or might still be better ways of getting the high-emitting less developed countries to join the industrial countries in making carbon cuts.

From the time climate negotiations began, those countries that have seen themselves as most endangered by climate change and least capable of coping with it—basically, the world's least developed countries—have put much more emphasis on obtaining help with climate adaptation than on emissions cuts as such. The tension between adaptation and gas reduction (or mitigation, as the official jargon has it) is another fundamental issue that will not go away. But it does show some signs of abating, at least in the current state of play.

When negotiations in preparation for the Rio Earth Summit began in 1990–91, there was an attempt to advance the principle of "polluter pays," which is widely honored in international environmental law. This is the commonsense notion that those suffering demonstrated damages deserve compensation from those causing the damage. It was expressly rejected as the Rio Framework Convention took shape. Many an international lawyer regretted that at the time and still does, but for now polluter pays is off the table and seems destined to return only if and when island countries start disappearing and low-lying coastal areas become uninhabitable.[12] Those scenarios represent the extreme cases, if you will, in the debate over how much emphasis there should be on climate change adaptation versus climate change prevention.

Initially, those most concerned about limiting climate change were rather suspicious of, even hostile toward, talk of adaptation, which they saw as a diversion from the main task of emissions reduction. There is still a good deal of suspicion even today. Naomi Oreskes, the coauthor of two important books about climate change, has likened the idea of climate adaptation to the quaint notion of "winnable nuclear war."[13] But that attitude is mostly obsolete. As the ill effects of climate change are becoming more visible to all and the uncomfortable truth has sunk in that climate change cannot simply be stopped in its tracks, climate activists have come to recognize and accept that adaptation assistance is inevitably an important part of any international climate effort. Representatives of developing countries, for their part, increasingly recognize that they do have a real stake in seeing emissions sharply reduced, and sooner rather than later.

The whole subject of adaptation could become much more controversial in the future again, as increasingly large sums are spent to address it and the question becomes whether "to protect as many dollars as possible, or to protect as many people as we can," as University of Massachusetts economist James K. Boyce has put it. Suppose, postulated Boyce in a simple thought experiment, world incomes were to plummet in the next decades as a result of drastic climate impacts. "Adaptation can cushion some but not all of these losses. What should be our priority: reduce losses for the farmworker or the baron?"[14]

And how much should be spent in total? Selwyn C. Hart, the secretary-general's point man on climate change, foresees a day when carbon cutting and climate adaptation will obtain about equal shares of the financial assistance available. How much financial assistance should be available? At Copenhagen in 2009 it was agreed that the rich countries should be providing one hundred billion dollars a year to developing countries. Although progress has been made in the meantime in establishing an office to manage the Green Climate Fund, pledges from the industrial countries are a far cry from approaching a hundred billion dollars, and there has been a lot of slippery talk about how much of the funding will be public sector and how much private. To the extent that it is private, how would one determine that it is truly additional to what would have been forthcoming anyway? To the extent that it is public, how is it to be rigorously distinguished from other forms of foreign aid?

We do not see a lot of politicians in the industrial countries going out on the campaign trail telling their constituents that they should spend much more money to help poor countries cope with climate change because it is their moral obligation. But when those political leaders or their diplomatic representatives are sitting in a room together and are seeking common ground, such moral obligations are taken for granted. The ultimate question is, What exactly are those obligations?

The field known as climate justice or climate ethics is surprisingly elaborate and has attracted high-power intellects with a rather wide range of attitudes and opinion.[15] A leading figure in the field, Stephen M. Gardiner, has memorably described it as "a perfect moral storm."[16] This is because, as he sees it, the causes and effects of global warming are widely disparate, the parties responsible are "fragmented," and the institutions needed to address the problem are inadequate to the task. On top of that there is an "intergenerational storm," because, in taking costly measures now to head off dangerous climate change, we are doing so primarily for the sake of future generations, for whom we are constantly sacrificing as it is and who will never be able to pay us back for our generosity.

Moral philosophers are in the business of thinking rigorously about individual responsibility, the responsibility of one person to another. But when we talk about our collective responsibility to members of future generations we do not know exactly who we are talking about. The membership of future generations will be different depending on what we decide to do about climate change right now, and if we do things that are too costly now, there may be fewer future people and they may be worse off. Indeed, at least one ethicist has alleged that seemingly irresponsible actions like continuing to spew greenhouse gases into the atmosphere uncontrollably may not actually harm many future people, because those actions could also be "necessary conditions of people coming into existence" in the first place.[17] That may seem a rather dubious point, but not necessarily to utilitarian philosophers who think of good and bad in terms of consequences, that is, the net balance of costs and benefits to individuals, over time.

The University of Chicago legal theorists Eric Posner and David Weisbach, in a book about climate justice, distinguish between "deontologists," who "focus on the rightness or wrongness of particular acts independent of their consequences, for example, certain acts such as lying," and "welfarists" (or utilitarians)—the camp they identify with.[18] Perhaps the best known of the utilitarians to write about climate change is the Princeton philosopher Peter Singer, who has made himself rather famous with down-to-earth arguments about how much the world's better-off individuals should spend to save the world's starving children. Applying similar reasoning to climate change, Singer has suggested that figuring out how to distribute the costs of addressing it is basically no different than dividing up an apple pie among friends and family: Figure how much smaller global greenhouse gases need to be by (say) 2050, distribute suitable emissions permits to all people in the world, and allow them to purchase and sell the permits freely—which will have the benign side effect of tending to equalize world income and wealth.[19]

An approach based on essentially the same philosophical premises would aim to equalize per capita greenhouse gas emissions globally while reducing them in a way fair to all. An influential version of this approach, dubbed "contraction and convergence," was developed in Britain during the early 1990s and has been elaborated by scholars associated with the United Kingdom's Tyndall Centre for Climate Change Research.[20]

The most obvious examples of deontologists (though they are not the ones Posner and Weisbach mention) are those taking the position that protection from global warming is an elementary human right. This philosophical position leads logically to the conclusion that dangerous global warming should be prevented regardless of what it costs: If a business is emitting noxious fumes that threaten the health and lives of neighbors, then it should be stopped.[21] This position, though rather extreme, has had some real diplomatic currency. In 2007, the Alliance of Small Island States, led by the Maldives, asked the UN Human Rights Council to take up the question of climate change. Subsequently, eighty-eight UN member states endorsed the council's recommendation that human rights expert bodies get involved in the United Nation's climate negotiating process.[22] In a decision that could have influence on other countries playing significant parts in negotiations, in June 2015, the Dutch high court ordered the government to reduce greenhouse gas emissions 25 percent from 1990 levels in the next five years rather than merely 17 percent; it grounded its ruling in human rights and environmental law, which it said were "mutually reinforcing."[23] Pope Francis's encyclical on climate change, issued the same month, likewise took a deontological position, arguing that humanity's obligation to protect God's creation is a given.

Generally, the notion that protection from climate change is a fundamental right and deserves absolute protection, whatever the cost, has been too absolutist to become a consensus position, and it is one that strikes utilitarians and welfarists as simply absurd. But Posner and Weisbach go to the opposite extreme. In their book on climate justice (a project Cass Sunstein was involved in until he joined President Barack Obama's White House staff), they argue that climate change is such a serious problem, the world must adopt the most efficient, cost-effective way of curtailing greenhouse gas emissions: "Climate change is sufficiently serious that reducing emissions at the lowest possible cost must be our central task." Climate change is so serious, indeed, that the job of addressing it should not be mixed up with issues of rich and poor. "As the risks from climate change increase, the problem becomes more severe: Sacrificing climate change goals for distributive benefits quickly becomes a bad trade-off if failure to reduce emissions leads to terrible consequences."[24] The task then is to identify the most economically efficient way of reducing greenhouse gas emissions adequately and somehow persuade all the countries of the world to adopt that method, even if it is not immediately advantageous for individual countries to do.

Utilitarianism, when rigorously or ruthlessly applied to individual moral conduct, can lead to absurdities. (Singer himself has conceded, for example, that there is no clear logical limit to how much rich persons should give to the poor; critics have complained that on his reasoning, there are circumstances in which an Auschwitz guard would be justified in continuing to work in that job.)[25] When applied collectively, welfarism may be susceptible to the same problems. Posner and Weisbach end up asserting that the world should simply agree on the most inexpensive and efficient way of achieving climate goals, without regard to the distribution of costs. This collides directly with the right to development. If all one cares about is how to reduce future levels of emissions most inexpensively, it obviously will be much cheaper for poor countries to stay poor and not emit more than for very rich countries to change the mode of life to which they have become accustomed—and that is a conclusion Posner and Weisbach do not shy from. "The treaty should stipulate the globally optimal abatement even though that is not what is optimal for poor states."[26] But with all due respect to them, even apart from considerations of elementary justice, it is simply fanciful to suppose that representatives of the poor countries, who after all are an overwhelming majority of the countries party to the Framework Convention, will agree to such a scheme. As John Rawls pointed out early in his seminal *Theory of Justice,* when people enter into a social compact, "no one has a reason to acquiesce in an enduring loss for himself in order to bring about a greater net balance of satisfaction."[27]

Of all philosophers, Rawls seems to get the most mentions in the literature of climate justice. This could be construed as more than a little odd, inasmuch as he never talked about climate change and rarely about environmental issues, and he expressly said that his theory of justice does not apply directly to relations among states. Also, he came out of what some in the world might see as a rather parochial Western tradition of thought, the social contract theorizing of Thomas Hobbes, John Locke, and Jean-Jacques Rousseau. The reason why Rawls is so often the point of reference when people gather to discuss climate ethics—whether their spiritual roots are Confucian, Buddhist, Hindu, Christian, Muslim, animist, or atheist—seems to be the centrality of fairness in his thinking. Fairness appears to be an idea that everybody readily grasps. And Rawls's particular formulation of fairness, though made in the somewhat artificial and awkward language of social contract theory, has proved to have a certain durability and depth: the idea that in making or evaluating a social compact, everybody should act as if behind a "veil of ignorance," oblivious to one's own position in the social or contractual order, so that even if some benefit more than others, the situation of the least fortunate is at least improved. "Since everyone's well-being depends upon a scheme of cooperation without which no one could have a satisfactory life, the division of advantages should be such as to draw forth the willing cooperation of everyone taking part in it, including those less well situated."[28]

The principle of fairness is deeply embedded in the 1992 Framework Convention on Climate Change in the form of what the treaty calls "common but differentiated responsibility," a principle so basic that it is often abbreviated in specialist literature as CBDR. What the treaty says, in article 3, paragraph 1, is this: "The parties should protect the climate system for the benefit of present and future generations of humankind, on the basis of equity and in accordance with their common but differentiated responsibilities and respective capabilities." Together with the treaty's objective of stabilizing greenhouse gases "at a level that would prevent dangerous anthropogenic interference with the climate system" (article 2), common but differentiated responsibility forms the bedrock of the Framework Convention. It is, as a recent authoritative assessment put it, "a universally accepted principle" among parties to the convention. So, "arguably, current obligations must be interpreted and future obligations structured in accordance with [its] basic tenets."[29]

That is not to say that the principle of common but differentiated responsibilities and capabilities is uncontroversial. Differentiated responsibility generally is taken to refer to the historic responsibility of the advanced industrial countries— most of them former colonial powers—for the greenhouse gases already stored in the atmosphere. Respective capability refers to the greater ability of richer countries to take costly and inconvenient measures to address climate change. You do not see political leaders in the industrial countries telling their publics that they need to take responsibility for the gas emissions of their ancestors, and you do not see them telling voters that they may have to take measures that could be costly in economic competition with the world's rising poor countries. But when those leaders go into international climate negotiations, they know—or they soon come to appreciate—that headway cannot be made unless the principle of common but differentiated responsibility is respected.

In the run-up to the December 2009 climate conference in Copenhagen, the chief US negotiator took issue with the notion of historic responsibility, or at least was seen to do so.[30] It did not go over well. When, early in the conference, a leaked draft treaty circulated that seemed to reject the principle of common but differentiated responsibility outright, the result was a spontaneous rebellion on the part of the developing country delegations. The industrial countries had to move fast to mollify them, and the result was the rather ambitious promise of providing one hundred billion dollars in climate aid per year by 2020—a pledge they show little sign so far of being able to keep.

The Rio convention, besides articulating a basic objective and fundamental principles, also included important gas reporting requirements and specified a long-term negotiating procedure. It called for annual meetings of parties to the convention, the so-called Conference of Parties, or COP. Some of those meetings have turned out to be much more important than others, and the Copenhagen meeting, COP 15, was one of the most important of all.

After Copenhagen and looking ahead to the next really significant meeting, COP 21 in Paris, in December 2015, negotiators sent mixed and complex signals about common but differentiated responsibilities. The agenda set for Paris at COP 17, in Durban, South Africa, made no mention of the principle. The Lima Call for Climate Action, adopted in Peru at COP 20 in 2014 to set the course to Paris, explicitly called for the 2015 agreement to reflect common but differentiated responsibility, but always "in light of different national circumstances." That language reflected a significant diplomatic objective of the United States, which was to blur the distinction between the industrial and developing countries and set the stage for the more advanced of the less developed countries to assume specific responsibilities. In the ungainly draft treaty that came out of Lima, which ran to more than one hundred pages, every time "common but differentiated responsibility" was mentioned, the phrase "evolving national circumstances" was added in brackets, no doubt at the behest of the US delegation.

Thus, the United Nations' Hart was wondering in mid-2015 whether the principle of common but differentiated responsibility might be reaffirmed but somehow reformulated at Paris.[31] The scholar Joyeeta Gupta, who is generally sympathetic to the positions and interests of the developing countries, has likewise conceded that some reformulation might be called for. In her most recent book she observed that differentiation is likely to be seen as fair only if "application is dynamic and allows countries to graduate in and out of responsibilities." It will not be seen as fair if it seems to give developing countries carte blanche to adopt and lock in carbon-intense technologies, with dire long-term consequences.[32]

When you consider that every single one of the 195 countries taking part in a Conference of Parties can press for additional or qualifying language to be inserted into a draft agreement, and when you consider the number and complexity of the basic issues that always will be in contention, it is not hard to see why six months before Paris the draft agreement still was running to more than eighty pages. Onlookers wondered, not for the first time, whether the diplomatic process has become too cumbersome and simply cannot work any more.

3

Can Diplomacy Deliver?

In the two and a half decades since serious international climate negotiations began, global greenhouse gas emissions and atmospheric concentrations have continued to rise, and at present the rates of increase are faster than ever. What is more, as Yale economist Nordhaus has observed, world carbon dioxide emissions were declining relative to world gross national product (GNP) at faster rates before global climate negotiations began than in the two and a half decades since Rio.[1] Admittedly, neither the 1992 Framework Convention nor the 1997 Kyoto Protocol specifically aimed to stop the rise in world emissions, and so on a literal-minded interpretation the treaties cannot be said to have failed in their objectives. But most of us are not literal-minded. The plain facts of the matter are that world emissions have been growing faster than anybody expected in the 1990s and that nobody really knows what to do about it. This has led to a general perception, across the political spectrum and among everybody from casual observers to the most authoritative specialists, that diplomacy is failing.

The Rio Framework Convention "sought to organize global action to address a threat to the global commons—the atmosphere and climate system on which all life on earth depends. Such global action, however, depends on national governments, whose first responsibility is to their own people and well-being. For that reason, the climate negotiations have faltered. Nations could not agree on who is to blame, on how to allocate emissions, or on projections for the future." So conclude Timothy E. Wirth and Thomas A. Daschle, influential former Democratic Party leaders in the US Congress.[2]

"The UN process has not worked because it involves too many countries and issues; it aims for progress too quickly. The result is a style of diplomacy that concentrates on getting agreement where agreement is possible rather than on crafting deals that actually make a difference," argues David Victor.[3]

Many other observers of the diplomatic scene are just as scathing. "The 1997 Kyoto Protocol . . . hasn't accomplished much. And subsequent climate conferences haven't come up with a solution," writes the Yale economics Nobelist Robert J. Shiller.[4] "The diplomacy of climate change appears stuck as ever," comments the *New York Times* economics columnist Eduardo Porter.[5] Five years after Copenhagen, where world leaders tried to "seal the deal" with a comprehensive new climate treaty, "there is little to show for it" and "perhaps the world community should take a different approach," suggests Yale professor of environmental law Daniel C. Esty.[6] "Behind the scenes, some are asking what happens if there isn't a deal in Paris [at the end of 2015]," says journalist Fred Pearce, "or even how much it matters whether there is such a deal."[7] "The annual UN climate summit . . . has started to seem less like a forum for serious negotiation than a very costly and high-carbon group therapy session," says Naomi Klein.[8] Joanna Depledge, a former staffer at the climate secretariat and now with the politics department at Cambridge University, concludes that the global climate change regime "has not only got 'stuck,' but is digging itself into ever deeper 'holes' of rancorous relationships, stagnating issues and stifled debates . . ., thus rendering itself unable to serve as a tool or arena for learning."[9]

Matthew J. Hoffman of the University of Toronto, a leading scholar of climate governance, has expressed doubt about "whether the multilateral process will ever be able to deliver the deep cuts in greenhouse gas emissions that the international scientific community warns are required to avert the most serious effects of climate change." Writing in 2011, Hoffman wondered whether the Kyoto Protocol represented the high point of such efforts and the onset of their demise.[10] Another top authority, Joyeeta Gupta of the University of Amsterdam, observed in her 2014 book that although negotiations were going forward, it seemed that the industrial countries were "unlikely to push the [climate] regime much further."[11] A third authority, law professor Cinnamon Carlarne of Ohio State University, said simply that she finds the subject "depressing" and doesn't like to talk about it.[12]

In essence, there are two main lines of critique: One is that the UN's climate negotiating process by nature cannot work, the other that the process has not been producing any significant results in practice and perhaps should be replaced by some other rather different diplomatic process. Ultimately neither position stands up to scrutiny, though, to be sure, there is a good deal to be said for both.

To take the second and more specific class of objections first, it simply is not the case that the UN negotiating process has produced no meaningful results. The 1997 Kyoto Protocol, though it fell short of its overall objectives, inspired the only important emissions cuts made by major emitting countries so far—including those made in recent years by the United States. (Without Europe's having taken the first big step, as described in the next chapter, it is highly improbable that the United States would have taken the second.) The 1992 Framework Convention established the Bonn climate secretariat, which has become an important technical authority and respected scorekeeper. The convention itself, laying out fundamental constitutional principles of the climate regime—the objective of preventing dangerous climate change, the definition of greenhouse gases that were to be tracked and limited, the principle of common but differentiated responsibilities and capabilities, the establishment of the secretariat, the yearly negotiating process, and sundry other detailed procedures—was a major achievement and is recognized as such by authorities across the spectrum. Arriving at a political and diplomatic consensus about the definition of dangerous climate change— the 2°C threshold—was significant as well.

Two outcomes of the negotiating process in particular are noteworthy: the action taken within their framework to reduce deforestation and improve agricultural practices, and the impetus the talks have given to the development of emissions trading systems around the world.

When it was first proposed in international climate negotiations that changes in land use having to do with agriculture and forestry be included in global greenhouse gas accounting, the idea was treated with considerable suspicion. A lot of people, including representatives of nongovernmental organizations (NGOs) and consulting experts, considered data on emissions from forestry and agriculture much less reliable than the information compiled from industry, power and heating, and transportation.[13] They also thought that changes in land use and forestry might be less durable. Representatives of less developed countries, and especially of heavily forested countries like Brazil and Indonesia, considered talk of land use change a potential threat to their national sovereignty.[14]

In the initial negotiations leading to the Rio Convention, Australia, Canada, Russia, and the United States pressed hard for land use change accounting in the agreement, while the Europeans initially aligned themselves with the developing countries, which largely opposed it. In the compromises reached in the Rio treaty and the follow-up Kyoto Protocol, inclusion of land use changes in greenhouse gas accounting won the day, in exchange for commitments from the industrial countries to agree on emission cuts. In due course, land use change accounting came to be a regular aspect of what the Rio Framework's climate secretariat in Bonn does.

In 2005, a group called the Coalition for Rainforest Nations proposed creation of a special program to prevent deforestation, which won the support of the UN Food and Agriculture Organization (FAO), the UN Development Programme, and UNEP. As a result of its initiative, a concept called REDD, for Reducing Emissions from Deforestation and Forest Degradation, was put on the official climate-negotiating agenda. The idea immediately came under sharp criticism, especially from representatives of indigenous peoples around the world, who worried that assigning a monetized value to the carbon contained in native forests would lead to traditional forest dwellers being squeezed out by corporate agricultural interests.[15] That led two years later at Bali, at one of the more important annual meetings of Framework Convention parties, to an elaborated version of REDD that was dubbed REDD+. It "rewards the governments of tropical forest countries with financial payments if they reduce their rates of deforestation and offer alternative economic opportunities for forest dwellers," as one student of the subject put it.[16]

REDD+ describes itself, a little awkwardly, as "an effort to create a financial value for the carbon stored in forests, offering incentives for developing countries to reduce emissions from forested lands and invest in low-carbon paths to sustainable development."[17] Pursuant to that end, it takes account of conservation, sustainable management of forests, and enhancement of forest carbon stocks. The general idea is to encourage all countries to prepare comprehensive national programs to conserve and enhance forests, with the United Nations and UN-affiliated organizations like the World Bank, the UNEP, and the FAO providing financial support and technical assistance. In practice, the emphasis is on adoption of standards and best practices.

Although it has not been crystal clear just how REDD+ intends to accomplish all its ambitious aims, the program has in fact inspired its sponsoring organizations and some states to step up to the plate with impressive initiatives. Within five years of its launch, some forty countries were developing plans to make themselves REDD-ready.[18] The World Bank launched a Forest Carbon Partnership Facility, which soon was supporting specific projects in fifty countries. Leading industrial countries, including the United States, the United Kingdom, and Norway, pledged $4.5 billion at Copenhagen to assess the total carbon value of the world's forests.[19] Financial support from Norway, its coffers rich with the proceeds of North Sea natural gas and oil, enabled Indonesia to put a two-year moratorium on conversion of its forests to other uses and did much to stimulate Brazil's efforts in the past decade. Since the 1990s, the rate of Amazonian deforestation decreased from about 2.7 million hectares (6.8 million acres) per year to about 1.5 million hectares (3.7 million acres) per year, according to Jeffrey Hayward of the Rainforest Alliance.[20] Since 2005, the Amazon deforestation rate has decreased 75 percent, said botanist Timothy Killeen of Agteca-Amazonica, in connection with a recent study he did with other researchers. He found that the Amazon is doing better at present than their best-case scenarios had anticipated.[21]

Administrative details for REDD+ were largely finalized in Bonn, in June 2015, at the final major preparatory meeting leading to Paris. This was widely described as the most important achievement of the meeting, "unexpected" and even a minor "breakthrough."[22] In a UN poll taken in seventy-five countries in conjunction with the Bonn meeting, action to protect tropical forests was found to be the second most popular means of addressing climate change, worldwide.

A closely related and about equally well liked way of tackling global warming, and one that emerged naturally from the REDD negotiating process, involves changing agricultural practices. Agriculture is a significant source of greenhouse gas emissions, and the introduction of larger-scale commercial practices can worsen matters.[23] At the same time, it is universally recognized that warming will have on balance adverse effects on the amount of arable land and on crop yields, especially in poor countries. So the major international development organizations—the FAO, World Bank, UNEP, and so on—are increasingly focused on helping farmers in the developing nations adapt to climate change and on encouraging governments to develop national plans. Since the practice of agricultural extension is a highly developed art, substantial returns from these efforts are to be expected.[24] Though the incorporation of concerns about climate change into the everyday activities of development organizations—"mainstreaming" as it is called in the professional literature—has limits, it is obviously as such a good thing.[25]

Net changes in greenhouse gas emissions associated with agriculture and forestry are now routinely incorporated in the important greenhouse gas inventories produced by the Bonn climate secretariat. Anybody consulting those inventories for a particular country can select either emissions without land use and forestry taken into account or with them included.

Another prominent issue that initially was hugely divisive in climate negotiations but is coming to be grudgingly accepted is greenhouse gas emissions trading. When trading first was suggested with some emphasis by the US delegation and its allies at Kyoto in 1997, it encountered fierce antagonism among the many environmental and nongovernmental organizations that had convened on the sidelines, as well as among most of the European countries and European Union. A provision encouraging the establishment of such markets was introduced into the Kyoto Protocol only as one of the last-minute compromises to get the Americans aboard.[26] Yet within fifteen years, a dozen greenhouse gas trading systems had started operation and a couple more were set to get going.[27] Among the more important were the European Union's Emissions Trading System (ETS), the US Northeast's Regional Greenhouse Gas Initiative (RGGI, or "Reggie"), California's big nascent carbon emissions market, and a half dozen pilot markets in China covering some sixty million people.[28] To a great extent, it was the Kyoto Protocol that gave legitimacy to the idea of greenhouse gas emissions trading and inspired countries and regions to set up trading systems.

Richard Sandor, a pioneer in emissions trading, has a telling story that shows just how quickly things moved, once

Kyoto gave trading its imprimatur. Several years ago, when he visited China to give talks on the subject, a host quietly told him that he needed to stop dumbing things down and raise the level of his game.[29] Evidently the Chinese did not need to have the basics explained.

An economist trained at the University of Minnesota in the Chicago school of free-market economics and an expert on midwestern futures markets, Sandor was an early proponent of emissions trading as a means of reducing pollution abatement costs. He is credited with having played a key role several decades ago in formulating and selling the idea of creating a sulfur dioxide trading system in the United States as a means of reducing acid rain. His basic idea was that if those who found it expensive to make cuts could purchase emissions permits from those able to make off-setting cuts at much less expense, then the overall cost of cuts would be greatly reduced. The acid rain trading system based on that concept, which came into effect in 1995, immediately proved to be a splendid success, producing more than $120 billion in estimated health benefits over the next two decades at a cost of just $3 billion—far lower than industry had feared and predicted.[30] Though the acid trading system got a lucky assist by the big shift from high-cost, high-sulfur Appalachian coal to low-cost, low-sulfur Powder Basin coal in Montana, Sandor says that it would be "dangerous" to assume that this shift would have occurred anyway, in the absence of the acid rain program.

Sandor talks the talk of bottom-up versus top-down, contrasting the market-based programs he prefers to top-down "command and control" like the US Environmental Protection Agency's coal pollution regulations or the government's automotive fuel-efficiency (CAFE) standards. In that spirit he engineered the creation of a private-sector, voluntary greenhouse gas trading organization, the Chicago Climate Exchange (CCX), and initially had considerable success signing up customers. Participating companies in the exchange wanted to gather experience with carbon trading, and for five years their numbers grew nicely, to total more than 330 by 2008. The prices set in that voluntary exchange got boosts in 2004 when the Kyoto Protocol took force, as well as with Europe's establishment of the mandatory Emissions Trading System. But the general expectation was that the United States also would create a cap-and-trade system, and when that failed to materialize in 2009–10, participation in the CCX dropped off sharply and the carbon price plummeted. Sandor's system promptly failed.[31] It turned out that without the top-down, you couldn't actually make the bottom-up work.

Simply put: The idea of greenhouse gas trading was inspired by the climate negotiations, and for it to work effectively, it depends on countries participating in those negotiations to set limits on carbon emissions, so that there will be a significant monetary cost—a "carbon price"—of making emissions.

A lot of the early skepticism about carbon trading could be attributed to several obvious factors. First, there was wide suspicion on the political left of any market mechanism, for the very reason that neoliberal economists liked markets so much. Second, Kyoto offered the devastated economies of the former Soviet Union and Eastern Europe, which were given relatively easy-to-meet emissions targets, the opportunity to make windfall profits by selling emissions permits to richer countries that wanted to delay cuts. Those profits were instantly dubbed "hot air" by critics, mainly among the nongovernmental organizations that were closely monitoring and substantially influencing the talks. Third, there was reason to fear that the caps established in cap-and-trade programs might be set too high and other offsets procedures be too lax, setting the stage for all manner of scams. When Europe's Emissions Trading System, the ETS, and the US Northeast's Regional Greenhouse Gas Initiative started up, regulators were in fact much too easy on industry and utilities, so that carbon prices were meaninglessly low, emitters had little or no incentive to adopt more efficient technologies or lower-carbon fuels, and there was a general appearance of money just being passed around from hand to hand, to no good end.

With the establishment of Kyoto's so-called Clean Development Mechanism (CDM), which entitles entities in less developed countries to obtain finance and sell offsets in connection with lower-emission projects, critics of trading found plenty more to complain about. In the first years of the CDM it became unequivocally clear that Chinese and Indian firms were concocting projects specifically to generate higher emissions so that they could then cancel or modify them, claim to reduce the emissions, and cash in.[32] Yet as offsets procedures and trading systems came under scrutiny, rules were tightened up and caps lowered. The main thing was to guarantee "additionality"—the principle that offsets would be permitted only when it could be shown that a project would yield emissions reductions that otherwise would not have occurred.[33] Developing countries came to appreciate what could be accomplished with legitimate CDM projects properly reviewed. Although there are those who believe that the Clean Development Mechanism is "looking moribund" and will require a big shot in the arm soon if it is to survive, the consensus among specialists is that it nevertheless became a constructive factor.[34] "These market-based mechanisms have evolved over time to become successful, with the CDM delivering millions of credits," concluded a leading scholar who has the basic interests of the developing countries at heart.[35] The CDM "has been remarkably robust," said another.[36] Clean Development Mechanism credits appear to have been an important stimulus to green energy projects in China and India: In its fifth global assessment of climate change, in 2013, the IPCC found "a clear contribution of the CDM to the rapid upswing of the renewable energy sector in China."[37]

In the late 1990s, when the Clean Development Mechanism was being organized and more ambitious trading systems were being conceived, Europe and the United States were still at loggerheads about this whole approach. As a result, the monitoring and verification of greenhouse gas offsets, instead of being delegated to the climate secretariat in Bonn, was turned over to private sector accountants. A system developed that was closely analogous to financial accounting, in which a few big private companies are responsible for most corporate auditing.[38] Emissions trading mechanisms have continued to arouse suspicion on the political left and among some environmentalists, as big financial players like JPMorgan Chase and Goldman Sachs got into what was becoming a multi-billion-dollar global business of trading.[39] But systems like the ETS, Reggie, and California's, after some growing pains, have come to work pretty smoothly from a technical point of view. Reggie seems destined to expand to include Pennsylvania and perhaps states of the upper Middle West, says Harvard's Robert N. Stavins, an authority on the subject; California's could come to include Oregon, Washington, and perhaps some Rocky Mountain states.[40] Governments of the world have largely come to embrace carbon trading systems, and they are no longer intensely controversial in diplomatic venues. The big bargain that Russia and other former Soviet states got is history, and so too are the missteps and foul play that attended the first trading and offsets programs. Thus, it is possible to imagine that in the long run, regional trading systems will be linked up to form nationally regulated markets, and that in the considerably longer run, the national markets might be integrated into a global system, so that there is a single recognized world price on carbon, the same way there is a recognized world price for oil.

Advocates of trading wondered if the new climate agreement adopted in Paris in December 2015 would do at least some little thing to nudge that process along. The hope was that as a new structure of collective carbon limitation takes form in the climate talks, first regional and then global carbon prices will emerge, which will vastly increase the overall efficiency of the world's greenhouse gas reduction program. Yet the Emissions Trading System carbon prices remained much too low to inspire big changes in energy investment plans, and the Clean Development Mechanism's future was gravely in doubt, as nongovernmental organizations continued to fight the whole idea of emissions trading and a system of voluntary rather than mandatory emissions cuts was taking shape. Without top-down mandated emissions reductions, it was questionable whether the CDM or any system like it could continue to work. Still, because so many emissions trading systems are being developed, they are sure to be an increasingly important part of the global effort to contain global warming, whether or not they get the green light in climate negotiations.

Not all issues relevant to climate change are addressable in the Framework Convention's diplomatic process, and it is not necessarily the most suitable venue for addressing certain issues. Anything that smacks of "putting a price on carbon" has always faced a sharp uphill battle in the UN climate negotiations, and any effort to mandate an end to fossil fuel subsidization is doomed to certain failure.

In keeping with general UN procedures and the accepted rules of sovereignty that date back to the 1648 Treaty of Westphalia, each and every country, however big or small, has in principle an equal say in the talks. All major decisions are adopted by consensus, and though the person chairing a conference of parties can make exceptions to that rule, this is done only on occasion and at some peril. One prominent example of an idea that has not been able to get onto the negotiating agenda as yet is the matter of carbon taxation, which is constantly being advanced by climate policy specialists and activists.

Among the economists concerned about climate change, from Jeffrey Sachs and Paul Krugman to William Nordhaus and Henry M. Paulson, there tends to be a distinct preference for carbon taxation over carbon trading.[41] Across the political spectrum, economists favor carbon taxation, and most people seeking strong climate policies agree (this writer included). One reason is that a carbon tax is much simpler than emissions trading: the US cap-and-trade bill that came a cropper in 2008–9, "Waxman-Markey," ran to hundreds of very complicated and highly contested pages; a bill taxing fossil fuels either at their source or at their point of use could in principle be just a few pages. Even more important, whereas a carbon trading system gives direct incentives and disincentives only to those players required to participate in the emissions market, a carbon tax gives a clear price signal to every single person living in the area affected by it. Higher prices at the pump for gasoline, higher electricity bills, higher home heating bills—all encourage consumers to reduce energy use and find lower-carbon or more efficient sources.

Because of those obvious examples, carbon taxes have been introduced in a number of countries and subnational jurisdictions, including Ireland, Sweden, Australia, and Canada's

British Columbia. Australia, for example, imposed a twenty-three-dollar-per-ton tax on carbon emissions from the nation's top five hundred emitters. Chile's president Michelle Bachelet boasted at the UN Climate Summit in September 2014 of having introduced a similar system that applies to large fossil power plants. International Monetary Fund economists have suggested that carbon taxes could, on balance, be beneficial to economies if the proceeds are used to reduce income taxation and improve public health. A Chinese carbon tax of sixty-three dollars per ton could cut the country's emissions by 17 percent, according to their calculations, while reducing the numbers sickened or killed by air pollution and boosting GDP 1 percent.[42]

Theoretically there could be an interconnected global system of carbon tax areas, with the tax graduated so that richer citizens in richer countries would pay higher fees than impoverished people in less developed countries. Yet carbon taxation, at least at this stage of the game, is almost always a nonstarter in American domestic politics and, by the same token, a diplomatic nonstarter as well. Ask anybody connected with US politics why emissions trading is so much more popular, and the usual answer will be simple: Find me an American politician who is willing to stand up and advocate anything that has the word "tax" in it. Kevin Rudd, a former Australian prime minister whose enlightened climate policies came to be repudiated by his electorate, suspects that the same political logic will prevent the word "tax" from appearing in global agreements as well.[43] Internationally, "is there *any* example of a globally harmonized tax system?" asked Daniel Bodansky, a top US climate negotiator during President Bill Clinton's administration and an important figure in technical discussions of climate diplomacy.[44] At the least, any attempt to put carbon taxation on the agenda of the yearly climate negotiations would immediately run up against the unalterable opposition of OPEC (Organization of the Petroleum Exporting Countries).

The OPEC countries have come under fire in recent years from the International Energy Agency (IEA), which has been drawing attention to the immense subsidies fossil fuels get in many parts of the world, totaling at least five hundred billion dollars per year.[45] The subsidies far exceed the amounts that go annually to the promotion of renewable energy sources, as the IEA has insistently reminded the world.

In principle, eliminating the subsidies could be a potent way of cutting carbon. A 2015 report sponsored by the Nordic Council of Ministers found that eliminating subsidies in twenty main countries would reduce emissions 11 percent on average and yield up to 2.8 billion metric tons in cumulative emissions cuts by 2020.[46] In practice, however, getting rid of the subsidies is an uphill battle. A common misperception is that the underlying problem can be reduced to the undue influence the major oil, gas, and coal producing companies have over government policy.[47] Were it that simple, one would just have to stand up to those energy behemoths, starting in the countries like the United States. But fossil subsidies in fact go mainly to consumers of oil, gas, or coal in fossil-rich countries. Direct US subsidies appear to be a small fraction of the global total.[48]

Though total subsidies and their shares are very hard to estimate authoritatively because of ambiguities as to just what a fossil subsidy actually is, about four fifths of total subsidies are believed to be going to consumers—probably four hundred or five hundred billion dollars per year. The intended beneficiaries are voters in countries like Iran, Russia, and Venezuela and the pampered though oppressed citizens of Saudi Arabia, Kuwait, and Qatar. Those are the countries that most firmly resist any attempt to introduce fossil fuel subsidization as a topic suitable for climate diplomacy.

Phasing out subsidies would cut world energy demand by nearly 6 percent, the equivalent of the energy consumed by Japan, Korea, Australia, and New Zealand combined. But this is a case where diplomacy cannot in fact deliver.[49] The chairperson of a Framework Convention Conference of Parties can in principle overrule a single obstructive country: at Cancún, Mexico, in 2010, for example, when it came to adoption of the conference's final decisions, Mexican foreign minister Patricia Espinosa simply dismissed objections from Bolivia, saying that one country could not be allowed to stand in the way of consensus.[50] But in practice, no chairman presiding over a conference of parties would try to override a major oil-producing country like Russia, observes Geoffrey Heal of Columbia University's business school, who has participated as an expert adviser in many a climate negotiation.[51]

So carbon taxation and fossil fuel subsidization are examples of important matters that cannot be addressed successfully in the context of the Framework Convention negotiations—or at least have not been so far—and such examples could be multiplied. But this would not prove that no important matters whatsoever can be negotiated in talks encompassing all the world's nations, or that the United Nations' climate-negotiating process is impossible as such. Although the Framework Convention and its negotiating process are virtually without equal in their universality, they are not without precedent. Since 1947, the nations of the world have been engaged in an increasingly all-embracing process of trade negotiation, first under the aegis of the General Agreement on Tariffs and Trade (GATT), then the World Trade Organization (WTO). Starting with twenty-three founding members, the GATT—actually a set of evolving rules, not a single agreement—came to have 102 participating members by 1973. The WTO, its successor, has upward of 159 members today. Though the trade negotiations have always been notoriously complex and controversial, through many "rounds" of negotiations (the Geneva Round, the Uruguay Round, and so on) countries have agreed to incur short-term disadvantages with respect to competitors for the sake of realizing a larger prosperity in the longer run. The Kennedy Round, for example, is said to have yielded, in thirty-seven months of negotiation, trade concessions amounting to forty billion dollars. Related negotiations like the North American Free Trade Agreement (NAFTA) and the Trans-Pacific Partnership (TPP) have been no less contested. And yet, however valid and serious the misgivings about various provisions may be, in a general sense the ongoing process of tariff reduction has come to be widely accepted across partisan lines as constructive.

Another ongoing negotiating process, admittedly not as sweeping as the trade talks but still important, is the procedure the nations of the world use to reallocate the radio spectrum every five years. In those meetings, sponsored by the International Telecommunications Union, delegations representing every country, from the big industrial states to tiny island nations like Tonga, quibble over every punctuation mark and every decimal point. Yet the results are sometimes dramatic breakthroughs. In the year 2000, when the iPhone was but a gleam in Steve Jobs's eye, the radio conference assembled in Istanbul freed up large amounts of spectrum for what was being called the "wireless Internet," setting the stage for the smart phone.[52] Very complicated and all-embracing talks can in fact produce important and useful results, and so, too, have the climate talks, contrary to so many of their critics.

Could climate diplomacy be more effective if it were confined to a much smaller set of negotiators, who more readily share a common lifeworld and represent countries with more tightly aligned national interests?[53] An alternative concept, which has come to be called club diplomacy, is not really so new but has gathered quite a bit of momentum in recent years as climate negotiations have faltered. Richard Elliot Benedick, chief US negotiator in the seminal ozone depletion negotiations of the late 1980s, put it like this: "The climate problem could be disaggregated into smaller, more manageable components with fewer participants—in effect, a search for partial solutions rather than a comprehensive global model. An architecture of parallel regimes, involving varying combinations of national and local governments, industry and civil society on different themes, could reinvigorate the climate negotiations by acknowledging the diverse interests and by expanding the scope of possible solutions."[54]

Echoing that view, Barry Blechman of the Brookings Institution and Ruth Greenspan Bell of the Woodrow Wilson International Center wrote in early 2014 that the better approach to climate diplomacy would be to "break the issues down and diversify negotiating venues and configurations. We need to take successive bites out of this huge challenge. A single global deal is too complex. Let's pursue smaller ones that have a reasonable chance of producing results."[55] Nordhaus, in early 2015, formalized the notion with what he called the club, a group of states that voluntarily share the benefits and costs of pursuing a common goal.[56]

Nordhaus took off from an important point he had made in *Climate Casino,* that an efficient and effective global regime to limit and reduce greenhouse gas emissions requires universal participation. On further technical analysis, he determined that any really ambitious effort to construct such a regime would open the door to "free riding"—the temptation on the part of major emitters to let everybody else do the costly carbon cutting while reaping competitive advantages for themselves. The free-riding loophole is so obvious, Nordhaus believes, that it is an almost insurmountable obstacle to adopting any carbon-cutting regime. So his proposal is that the major emitters form a club and agree to adopt a common target price on carbon—in effect, a common carbon tax—and then impose trade penalties on any major emitter that refuses to join the club and go along with the program.

Nordhaus seems to think that if the global carbon price is set just right, almost all countries will participate and few would be penalized. Nevertheless, he emphasizes that his club procedure should be adopted only after careful consideration because of the threat it would pose to free trade. "Today's open trading system is the result of decades of negotiation to combat protectionism. . . . A regime that ties a climate-change agreement to the trading system should be constructed only if the benefits to slowing climate change are clear and the dangers to the trading system are worth the benefits."[57]

Just as important, arguably, are the dangers that Nordhaus's trade-sanctions club would pose to the climate negotiating system itself. Imposition of trade penalties on free riders inevitably would lead to their ejection from the Framework Convention negotiating process and could result in its complete unraveling. Running that risk certainly has been considered by important players in the talks in the past, but so far it always has been rejected. During the first decade following adoption of the Kyoto Protocol, the critically important free rider was the United States, which refused to ratify it and then repudiated it. The Europeans, who were taking strong measures to meet their commitments, must have been sorely tempted to penalize the United States but refrained from doing so to protect the larger diplomatic process. In the next decade, when China's emissions were swamping everybody else's, the temptation of resorting to sanctions reared its head again, and again the impulse was resisted.

As faith in the UN negotiating process has wavered in the past fifteen to twenty years, first because of Kyoto and then because of the disappointing and disconcerting 2009 Copenhagen conference, a great many important initiatives have been taken outside of and parallel to the Framework Convention negotiations. In 2009, the G20 (Group of Twenty) countries agreed to "phase out and rationalize . . . inefficient fuel subsidies" and to form a system of surveillance so that they would nudge one another along.[58] Major cities around the world, individually and collectively, have adopted carbon reduction commitments and planned programs to execute them. The same has gone for regions, provinces, and states around the world. Ultimately, however, all those initiatives in aggregate will accomplish the mission of preventing dangerous climate change only if there is a firm and formal global consensus that they must accomplish that objective. That consensus can be expressed only in the context of the UN talks.

Ten years ago, when California adopted one of the world's most far-reaching and ambitious climate and energy action programs, its then governor, Arnold Schwarzenegger, memorably took note of the nightmarish scenario in which the vitally needed Sierra snowpack would be lost. Climate change "creeps up on you," he said. "We don't want to go there."[59] Well today, ten years later, the snowpack is greatly reduced and California is suffering its worst drought in a hundred years— one whose effects have been seriously aggravated by global warming. The citizens of California are learning (if they did not appreciate it already) that climate change cannot simply be stopped in its tracks by unilateral action and that in the not too long run things are almost sure to get even worse, unless somehow their actions to rein in climate change are matched by comparable actions everywhere else in the world.

Programs like California's not only depend on the global collective process for success but also, to a great extent, have been inspired by it. As Depledge has observed, "The launch of the European organizations, and the emergence of a variety of initiatives by municipal and state governments, including in the US—all have been stimulated, directly or indirectly, by [the climate change] regime."[60]

It may be that the United Nations' climate-negotiating process seems just too cumbersome and just too complicated. But arguably, the shortcomings of the Rio Convention's climate negotiating process are not the result of flaws intrinsic to and unique to that particular process. Rather, they are to be laid at the door of the major participants in that process, none of whom escapes serious criticism on close review, though some have behaved much more constructively than others. That, anyway, is the thesis of this book.

II

The Players

In general, situations appropriate for negotiations have two characteristics: the parties agree they need a solution ("we can't go on like this"), and that their decision on a solution must be unanimous ("we're all in this together").

—I. WILLIAM ZARTMAN AND MAUREEN R. BERMAN, *The Practical Negotiator* (1982)

It is dangerous to carry distrust of professional diplomacy to the point where you always insist upon doing what the professionals say must not be done and always refuse to do what they describe as necessary.

—GORDON A. CRAIG, *The Diplomats, 1919–1939* (1967)

4

The Superpowers

From the beginning of climate diplomacy twenty-five years ago, the world's two economic superpowers, the European Union and the United States, were almost never on the same wavelength. That situation was all the more unfortunate inasmuch as the two overwhelmingly command the most intellectual resources bearing on the problem: not only in climate science itself, but in all the fields bearing on the prevention of dangerous climate change, from economics to computer modeling. Their influence on how issues are formulated and addressed in negotiations is disproportionate, even taking their immense economic clout into account.[1] Though diplomatic differences between the two major parties narrowed some in recent years, contrasts between the two still outweighed commonalities until very recently.

From the initial negotiations that led to the adoption of the Rio Framework Convention in 1992, Europe has consistently advocated mandatory greenhouse gas emission cuts on the part of the advanced industrial countries, and it has consistently acted in concert to reduce its own emissions. Considering that the twenty-eight members of the European Union represent the world's largest economic bloc—their combined gross domestic product exceeds the US GDP by about one trillion dollars—their unwavering support for strong climate action is no small matter.

The Framework Convention, among other important things, provided that parties would develop and submit to the treaty's secretariat national inventories of emissions, taking into account removals or "sinks"—negative emissions if you will. So, since the early 1990s, the secretariat has compiled that information, acting in effect as the world's scorekeeper. The European numbers are dramatic. From 1990, the baseline established in the Rio treaty and its 1997 Kyoto Protocol, to 2012 (the final year in which initial Kyoto commitments applied), emissions of the fifteen European countries that initially signed the treaty decreased 17 percent—more than double what the protocol required of them. Emissions of the twenty-eight countries that are now members of the EU decreased 21 percent.[2]

The Kyoto Protocol, which committed the advanced industrial countries to targets that were to be achieved by 2008–12, allowed for wide variation among the European states.[3] Spain, for example, was permitted a 15 percent increase—and ended up exceeding it, registering 18 percent in 2012. Sharp decreases were expected of the United Kingdom and Germany, on the other hand, and both countries outdid their targets by wide margins. British emissions were down 27 percent in 2012 from 1990, and Germany's were down 23 percent. Cuts made by the Scandinavian countries also exceeded requirements. Denmark's, for example, were 37 percent lower.

From the beginnings of climate diplomacy in the first years of the 1990s, Britain and Germany have been the European leaders. They generally have had the unwavering support of the Scandinavians and of France—all rather green countries to begin with, in terms of greenhouse gas emissions, mainly because of their very low carbon electric power sectors. Both Germany and the United Kingdom benefited going into the Rio-Kyoto negotiations from existing policies that eased the way to carbon cutting: Under Thatcher, the United Kingdom had opted to phase out reliance on coal, which had been the country's largest source of greenhouse gas emissions; Germany, following reunification, planned the reorganization and rationalization of the east's archaic electric power system and industrial infrastructure, which were almost wholly dependent on lignite, the dirtiest form of dirty coal. Both countries adopted ambitious policies following Kyoto that consolidated gains to be expected from deemphasizing coal and set green goals that went far beyond what would have been automatically gained from the coal policies alone.

At the end of the 1990s, Germany, governed at that time by a coalition of its Social Democratic Party and the Greens, adopted a so-called feed-in electricity tariff. Anybody or any entity producing renewable electricity—whether from wind, solar, or some other nonfinite resource—would be guaranteed a set price for each kilowatt-hour produced, on a time schedule set in the law. The considerable costs of these renewable energy production subsidies would be distributed among all German ratepayers, so that everybody's monthly electricity bill would go up some fraction to support the program. The feed-in tariff proved to be a powerful policy instrument and, together with Germany's traditions of technological and industrial prowess, quickly made the country the world leader in green energy innovation and implementation. The law was widely copied in other parts of Europe and indeed globally, sometimes too uncritically. Spain, for example, got itself into serious trouble with solar subsidies that were much more generous than it could afford. The subsidies were not completely uncontroversial in Germany itself. By the time of the 2013 election, when Chancellor Angela Merkel was running for a third term, high electricity rates threatened to become a significant campaign issue.[4] Merkel had held fast to the "energy transition" policies inaugurated by the predecessor "red-green" coalition government, though her job was complicated somewhat by a decision to phase out Germany's reliance on nuclear power, following the Fukushima catastrophe.[5] At least in the short term, getting rid of the zero-carbon reactors meant more reliance on coal-fired generators.

Britain's current climate and energy policies were formulated in a 2007 White Paper, "Meeting the Energy Challenge," and in the 2008 Climate Change Act.[6] They called for cutting the United Kingdom's greenhouse gas emissions 80 percent by 2050, from their 1990 level, while at the same time guaranteeing all citizens reliable and affordable electricity and heating. That implies installing 30–35 gigawatts of new electrical capacity—the equivalent of thirty to thirty-five full-scale nuclear power plants—to replace coal-fired plants and reactors being decommissioned. To meet that immense technical and economic challenge, the Climate Change Act called on the government to prepare future carbon budgets, rely on the advisory Committee on Climate Change, and prepare a national climate adaptation plan.

Again, aspects of the country's climate and energy policies have not gone without controversy. The conservative government currently in power has sought as part of its climate plan to embark on a new generation of nuclear construction but ran into opposition from environmentalists, led by Greenpeace International, which stopped the program in its tracks with a legal intervention. Construction of wind farms has aroused passionate antagonisms among rural English, who love their gardens and birds. Organizations like Friends of the Earth, which cut their teeth in the 1970s and 1980s organizing local people to *oppose* construction of nuclear power plants, now expend much of their energy *persuading* locals to allow construction of wind farms. Partly because of this controversy, the British government embarked early in the last decade on a very ambitious program of offshore wind construction, in the unruly waters of the North Sea and Irish Sea.

Anybody who says that countries like the United Kingdom and Germany have adopted ambitious green policies merely because they found it so easy and inexpensive is simply wrong. But why is it that the countries of northwest Europe—not just Germany and the United Kingdom but also Scandinavia and the Low Countries—have been so steadfast in their climate commitments? European publics have pretty consistently been more sharply concerned about global warming than the American public has been, though, to be sure, opinion has tended to fluctuate in similar cycles: Climate concern has abated somewhat in times of economic duress while picking up again in more prosperous times.[7]

Two scholars of climate policy contrasted American and European attitudes as follows in a 2010 paper: "According to a recent study by the German Marshall Fund, while 65 percent of Americans are 'worried' about climate change (compared to 84 percent of Europeans), only 43 percent of Americans are willing to sacrifice economically to slow global warming, compared to 69 percent of Europeans. . . . While Europeans ranked climate change as one of the world's most serious problems (above international terrorism and a major global economic downturn), even among Democrats in the United States it ranked below health care, education, social security, the budget deficit, and illegal immigration."[8]

It is not obvious why Europeans are so engaged in the matter of climate change, and since it is not clear where their sentiments originate, it also is not clear how deep they go. Possibly, considering that every European with a grade-school education knows that the climate of northwest Europe is an anomaly—that their civilization would never have flourished if ambient temperatures were those of Canada's Northwest Territories, directly to the west—Europeans may take it for granted that any significant change to the global climate system could be very bad news for them.[9] But if this is the case, it does not show up in opinion polling, which rarely drills down enough to tap into this kind of consideration.

In the last analysis, ironically, Europe's climate constancy may have less to do with public opinion and more to do with what often goes by the name of its "democratic deficit." Climate policy is largely the creature of the European Union, which itself is governed by the twenty-eight heads of state (or their representatives) serving on the governing European Council. And those heads of state—from Thatcher, an Oxford-trained chemist, to Merkel, a PhD physical chemist—have generally "got it" when it comes to climate science and climate policy. But there are institutional, legal, and political considerations that also account for why those leaders have been able to largely set policy.

Throughout western Europe, the reaction to the defeat of fascism in World War II was to establish social democracies, in which the welfare state and some degree of economic and social planning are taken for granted.[10] Except in England, where Thatcher challenged that underlying consensus in the 1980s and 1990s, it is generally assumed in the countries of western and northwestern Europe that the costs of needed national policies will be fairly distributed among social classes and stakeholder groups through a process of internal bargaining. This general approach came to be most highly developed in West Germany and to be known as the Rhineland Model, referring to Germany's industrial region of the lower Rhine, the Ruhrgebiet. Within that model, it is a relatively straightforward matter to decide how the social and economic costs of a climate policy will be shared—and in this narrow sense Naomi Klein was right when she asserted that effective climate policy is at odds with the neoliberal, deregulatory policies of Thatcher and Reagan.

Add to that the general modus operandi of the European Union, in which European climate policy is largely set. Here, too, it is taken for granted that the costs of any policy adopted will be distributed among the member states, based you might say on their respective capabilities and historical responsibilities, through a process of bargaining among sovereign heads of state. In the United States, by contrast, as law professor Cinnamon Carlarne has pointed out, it is actually unconstitutional to make laws that have specifically differential impacts and costs in different parts of the country.[11] To put matters in the simplest possible terms, Congress cannot enact, for example, a national carbon tax but then rebate the proceeds specifically to the traditional coal-producing states—say, West Virginia, Pennsylvania, Kentucky, Ohio, and Indiana. In Europe that would be not merely allowable but routine.

There can be downsides, of course, to the way Europe constantly takes pains to accommodate major interest groups. During the first years of the Emissions Trading System, allowances were much too generous and carbon prices much too low largely because of lobbying on the part of the continent's immensely influential steel and auto industries. In the ozone diplomacy that led to the Montreal Protocol and laid the foundation for Rio (chapter 7), industry's influence made Europe a relative foot dragger (though that later would change).[12] Generally, though, the European practice of appeasing interest groups and stakeholders has been a major positive factor, enabling it to advocate and introduce potentially costly climate policies.

European Union decision-making procedures do give rise to a lot of bickering among member states over climate policy in Brussels. The annual rotation of the European Council's presidency can lead to awkward situations, too, as Carlarne has noted.[13] For example, in 2009, a crucial year in climate diplomacy, the Czech Republic's president Vaclav Klaus became the European Union's titular leader. Klaus, a free-market economist, had written a book in which he declared environmentalism an enemy of human freedom. On the other hand, as Carlarne also notes, there is the obvious fact that daily politics and politicking are generally much less contentious in Europe than in the United States.[14] The climate issue has *not* come to be politicized along left-right lines the way it has in North America, and leaders do not have to worry that if they take certain kinds of forward positions on climate that they automatically will come under sharp attack from a big exposed flank.

As a result, Europe's leaders have consistently sought to be an example to the rest of the world in climate policy. Among scholars of the subject, they are credited with having done so while avoiding open conflict with less ambitious major powers, notably the United States, and have taken great pains to keep them on board in negotiations. The negative side of this, perhaps, is that they tried *too* hard to avoid open conflict with less ambitious major powers like the United States, regardless of the costs in terms of climate policy effectiveness.[15]

Thus, whereas Europe played a key role in the adoption of the Kyoto Protocol and the 2°C standard at Copenhagen, it was not able to maneuver the United States back into the

Kyoto regime and has had to accept an international program of voluntary national emissions cutting that may or may not yield adequate rewards in the long run. Its moral influence does have real scope: Not only in the United States but around the world, some regions and municipalities have taken Europe as a reference point in adopting climate programs. But Europe's moral influence has limits, and what is singularly distressing, those limits seem to get little systematic discussion among its opinion leaders. Typical was a *Spiegel* magazine article about climate change and climate policy that appeared in February 2015. Geared specifically to the upcoming Paris conference and running more than five thousand words, it had not a single word addressing the glaring question of what German leaders could do to better achieve their most important aims in diplomatic negotiations.[16]

Simon Schunz, a professor at the College of Europe, put it like this in 2011: "The [European] Union simply has failed to obtain most of its substantive [climate] policy objectives, not just in Copenhagen, but also in previous rounds of talks." There has been "a striking discrepancy between its undeniable long-term engagement in global climate politics and its track record of attempted, but failed leadership by example."[17]

Admittedly, even if the Europeans were willing to play much harder ball in climate negotiations than they have been to date, it is not obvious that they would accomplish more. As one student of the subject has observed, though their leadership in climate policy has translated into "little actual problem solving," they may not be able to even sustain that leadership, let alone become more effective, because of "ongoing shifts in the geopolitical landscape."[18] An inconvenient diplomatic truth, as David Victor observed trenchantly more than a decade ago, is that the more a party to the Framework Convention succeeds in cutting its emissions, the smaller its influence becomes in the ongoing negotiations.[19] Conversely, those contributing most to making global warming worse at any given time are likely to have the greatest influence. Thus, looking ahead, the largest emitters among the developing countries have the most bargaining power, since they are the baddest boys on the block. And looking back to the decade following adoption of the Kyoto Protocol, it was the United States that was the free rider, demanding changes in the basic rules of the game while benefiting from the emissions cuts made by the very parties it was berating.

The Kyoto Protocol had required the European Union to cut its emissions 8 percent by 2008–12, vis-à-vis 1990, and the United States to cut its by 7 percent. Eight years after adoption of the protocol, in 2005, US emissions were about 15 percent above their 1990 level, while emissions of the EU-28 were down more than 9 percent and of the EU-15 more than 2 percent. By then, US policy spokespersons were well entrenched in the position that the European cuts did not really count because they had been so much easier to make than US cuts would have been (an issue we shall return to in chapter 8). The more fundamental and enduring reason for US obstructionism was the growing economic might of China and, closely related to that, its fast-growing emissions. As it happened, it was in the late 1990s that China suddenly was seen to be coming on strong as an economic powerhouse, and concern about adverse impacts on US industrial jobs was at a peak in the United States.

In July 1997, months before the Kyoto Protocol was concluded, the US Senate resolved by a vote of 95–0 that it would not ratify any climate treaty that failed to require developing countries to limit their emissions or that put the United

States at an economic disadvantage.[20] The sponsors of the resolution were two major figures: the late senator Robert Byrd of West Virginia, the Democratic Party's seniormost elder statesman at that time, and Republican senator Charles Hagel of Nebraska, who would later serve as secretary of defense under President Barack Obama. Despite the Senate's near-unanimous support of the resolution, the US delegation proceeded, largely at the behest of Vice President Al Gore and climate lobbyists, to agree to the protocol committing the United States to a 7 percent reduction in emissions by 2008–12.[21] It is at this point that US climate policy became, by comparison with that of the European Union, almost wildly erratic. President Bill Clinton affirmed that he would act to fulfill his Kyoto pledge, regardless of the Senate's attitude, but when President George W. Bush succeeded him and took office in January 2001, he promptly made it known that the United States was repudiating the protocol and would not try to meet its provisions.

There followed an eight-year period that was generally seen by climate activists, both in the United States and worldwide, as a dark age in climate policy. Though the United States continued to participate in all-important global climate negotiations, its role was not generally constructive and its presence tolerated only because it was much too important to throw out. Its emissions continued to rise until 2006–7, when, not because of any deliberate policy, the sudden boom in fracking led to the substitution of natural gas for coal in electric power generation. (Per unit of electricity generated, gas produces only a third or a half as much carbon emissions as coal.) In the 2008 presidential campaign, as in 2004 and 2000, candidates did not go out of their way to address climate change.

Nevertheless, on his election in November 2008, President Obama promptly made it known during the transition that he was going to make a radical new beginning in climate policy and that from now on, anybody seeking to develop green or clean energy would have "a friend in the White House."[22] It was generally and naively assumed that this signaled that the United States would now get back into the mainstream in climate negotiations. The year before, at a relatively important Conference of Parties (COP 13) held in Bali, an "action plan" had called for agreement to be reached in 2009 on a second commitment period under the Kyoto Protocol, covering the years 2012–20, and on a new climate treaty to succeed Kyoto starting in 2020. But at Copenhagen in December 2009 (chapter 9), the Obama administration again expressly rejected the Kyoto approach, insisted on emissions limitations on the part of the major emitting less-developed countries, and rammed through an informal "accord" calling on all countries of the world to submit climate pledges that would then be independently assessed.

Though that outcome blindsided foreign partners of the United States and caught most climate activists completely off guard, in hindsight it is evident that the swings in US policy were never quite so extreme as they appeared on the surface. The United States consistently took the position, for example, that the bigger emitters among the developing countries should be required to set limits; President Bill Clinton's administration hued to that line in a 1996 COP in Berlin, where the groundwork for Kyoto was laid.[23] As for George W. Bush's administration, there was a telling moment on July 11, 2007, when Bush's senior climate negotiator Harlan L. Watson was testifying before a House subcommittee. When Republican representative Dana Rohrabacher lobbed him a softball question, obviously expecting him to say that all the talk of climate change was a bunch of hooey, what Watson said was: "The reality is we are changing the chemistry of the climate . . . primarily by burning fossil fuels.

... There is also no doubt about the measurement that the Earth has warmed approximately 1 degree Centigrade over the last 100 years."[24] Six months later at Bali, Watson told Surya Sethi, a member of the Indian delegation, that if other countries had been finding the Bush administration tough to deal with, they had better get ready for the Senate Democrats from the coal-burning states and the Rust Belt, who were most threatened by foreign competition and would be even rougher.[25] At Bali, the United States floated the idea of creating a climate regime based on voluntary bottom-up pledges rather than top-down mandatory cuts.

The rejection of the whole Kyoto process by the Obama administration at Copenhagen, and its insistence instead on an all-inclusive process of "pledge and review," was a drastic, almost violent act. In effect, it tossed out most of what had been going on during the previous fifteen years of climate diplomacy and restarted the process where it had stood in 1991–92, when Japan had advocated a pledge-and-review system, with the support of the United States and some other major industrial countries; that approach had been rejected at Rio and Kyoto by a coalition of Europe and the developing countries, which insisted on mandatory emissions cuts on the part of the industrial nations. By 2009, the Kyoto approach had come to be almost universally dismissed as a failure in US policy-making circles, not mainly because the United States had refused to participate in it, but because sharply rising greenhouse gas emissions in China were swamping the cuts made by Europe and others. For that reason, what the Obama administration did at Copenhagen was pretty generally seen in the United States as a necessary deed, however nasty it appeared.

Harvard's Robert N. Stavins eloquently captured that position with this metaphor: He said that with Kyoto, it was as if one were trying to build a huge skyscraper starting from a tiny foundation. Anybody looking at the construction under way could see that the building would be unstable and fail. To succeed, argued Stavins, it was necessary to start completely over from a much bigger foundation, and this exactly is what the Obama administration got the rest of the world to agree on at Copenhagen.[26] The United States had advocated requiring emissions commitments from the developing countries going back to the Rio and Kyoto negotiations, and now at last it had succeeded in getting the developing countries to agree to that in principle.[27]

At Copenhagen, let it be said, Obama did commit the United States to a 17.5 percent cut in its greenhouse gas emissions by 2020, leaving a wide, though mistaken impression among Americans that the United States was in effect meeting or even exceeding its Kyoto goals.[28] (In fact, 17 percent reduced emissions by 2020 would still leave emissions only 4.5 percent below their 1990 level, 3.5 percentage points short of what Kyoto would have required the United States to do eight years sooner, by 2012.) Still, over the six years following Copenhagen, somewhat improbably perhaps, the United States gradually won the world's general assent to the pledge-and-review system initiated at Copenhagen. That was mainly because, in the end, President Obama put his money where his mouth was.

During his first term, though Obama took some important clean energy measures—he sharply boosted automotive fuel efficiency standards and included large green energy subsidies in the 2009 economic stimulus bill—he largely kept his head down. The refusal of the Senate to act in 2010 on a cap-and-trade bill that had passed the House sent a clear signal about the general state of US opinion. In deference to public attitudes and industry lobbying, the Obama administration actively resisted European efforts to make international aviation subject to the European Union's Emissions Trading System, even enlisting Russia and India. Because of those efforts, the Europeans had to retreat from their efforts to require foreign carriers to purchase carbon emissions credits, which might have boosted ticket prices by about forty-five dollars.[29] Obama personally stayed mum on the subject of climate change during his first term and during his 2012 reelection campaign.

But immediately on reelection Obama came out of the closet. On June 25, 2013, in an eloquent speech delivered outside on a sweltering summer day at Georgetown University, in Washington, DC, Obama unveiled an extremely ambitious and comprehensive national climate action plan.[30] Major elements would include strict new air pollution regulations requiring sharp carbon cuts by the major coal-fired generators, still stronger emissions standards for both cars and trucks, support for all zero-carbon and low-carbon sources of energy, including nuclear power plants and fracked natural gas, and a wide variety of measures to encourage energy conservation and efficiency. Obama had signaled earlier that if Congress continued to reject cap-and-trade legislation, he would use all the regulatory powers at his disposal to bring down greenhouse gas emissions, and now he did so. Emissions already had dropped more than 10 percent from a peak in 2005 to 2012, largely as a result of the 2008–9 financial collapse and the introduction of fracked natural gas. But now, because of Obama's ringing speech and action plan, they began to decrease for policy-driven reasons as well, as critics of US policy like India's Sethi have conceded. "Now, because of Obama's personal commitment and efforts, policy-driven measures are beginning to bear fruit. However, there is still concern that consumption in the US and other rich nations continues to grow" and that consumption-based emissions are still rising in the top 20 of the world's countries, Sethi observed.[31]

By the middle of Obama's second term, pollsters were also registering a distinct change in US public opinion on climate change. Concern had first peaked in 2006, with prominent cover features appearing in diverse magazines like *Vanity Fair, Wired,* and *Time,* which showed a forlorn polar bear floating on a melting ice sheet.[32] With the onset of the economic crisis, concern dropped sharply (and not only in the United States). By the time the economy started to stabilize and recover several years later, concern about climate started to rise again. The country's experience with drastic climatic events—the Midwest drought of 2010–11, the California drought that began in 2013, Superstorm Sandy, and Hurricane Katrina—must all have had some cumulative impact on public perceptions. Now, strong majorities of Americans were expressing serious concern about climate change, and though the issue rarely ranked high on anybody's list of concerns, majorities were indicating that they would punish political candidates who opposed action on climate change and reward candidates favoring climate action.[33] Notably, though Americans wanted to see stronger action on climate, they also expressed a high degree of skepticism about whether government action could be effective. Possibly that just reflected a general and somewhat irrational antagonism to the federal government, but possibly it also had to do with a perfectly rational skepticism about whether the US government would be able to secure a strong enough international agreement to make the country safe from more disastrous climate changes.

However that may be, as the Republican and Democratic presidential campaigns were ramping up in 2015, it was clear that for the first time climate change would enter into the 2016 presidential contest: The Republican candidate would take issue with President Obama's aggressive climate action program, while the Democratic candidate would promise to continue with Obama's policies and, presumably, would be called on to defend whatever climate agreement Obama secured in Paris, in December 2015. Speaking at Yale University's law school on October 15, 2015, chief US climate negotiator Todd Stern said he expected climate denial to be a political liability in the 2016 presidential election. By then, that was coming to be common wisdom, at least among Democratic Party strategists.[34]

An anomalous feature of US politics was the sharp partisan divide that had developed over climate action. Though significant minorities of Republican voters expressed concern about global warming, the party's political leaders, virtually without exception, promised to slow climate action, whereas Democratic Party leaders unanimously promised to keep it up. In this respect Europe remained deeply different. There, concern about climate change was felt across the political spectrum. Except in Eastern Europe, where an anticommunist phobia about anything reeking of central planning was manifest, almost all mainstream political leaders were committed to continuing strong action on climate.

Yet despite that contrast, the positions of the United States and the European Union appeared to be narrowing as the December 2015 Paris conference approached, though significant differences remained. In his Georgetown University speech, Obama said that his administration was redoubling efforts to obtain a strong agreement and called for an agreement that would be "ambitious," "inclusive" (that is, universal), and "flexible." He said that he had no patience with those who denied the reality of climate change and expressed the belief that

America's founders visualized its political leaders as "elected not just to serve as custodians of the present, but as caretakers of the future." So, he said, "That's what the American people expect. That's what they deserve. And someday, our children, and our children's children, will look at us in the eye and they'll ask us, did we do all that we could when we had the chance to deal with this problem and leave them a cleaner, safer, more stable world? And I want to be able to say, yes, we did."[35]

Tellingly, Obama did *not* call for a legally binding Paris agreement, though the agreed-upon action plan for Paris specifically requires an agreement "with legal force." His administration naturally did not want to enter into yet another climate treaty only to see the Senate reject it. Though the Europeans were well aware of this dilemma, the question of what legal force the Paris agreement should have remained the main point on which they still begged to differ. Speaking at the UN Climate Summit in September 2014, Britain's David Cameron said that the Paris agreement "must be legally binding so we can hold each other to account." Chancellor Merkel did not attend the summit. (Was she sulking in her tent?)[36] The website of Germany's foreign office said that the European Union had been charged with taking a more aggressive stance in diplomacy, at the behest of the United Kingdom and Germany. It said that Germany's main goal at Paris would be to push for ambitious international emissions targets.

As for France, the host of the upcoming December 2015 conference, the Quai d'Orsay's website said that it, too, wanted "the EU to adopt ambitious emissions reduction objectives and to strengthen its leadership in negotiations." It said that France, with its relatively low and decreasing per capita emissions, would express national "exemplarity" at Paris and "engage in real climate diplomacy." At the beginning of 2014, French president François Hollande, Merkel, and Cameron prevailed on the Europeans to commit themselves to a 40 percent reduction in the European Union's collective greenhouse gas emissions by 2030, which would keep them well in the forefront of those seeking to limit emissions, still far ahead in that respect of the world's two top emitters, China and the United States.

A former top US climate diplomat, Nigel Purvis, took the position after Copenhagen that closer and better US-European cooperation on climate policy was now the sine qua non of any effective climate agreement. Writing with Andrew Stevenson in 2010, he said that future success would depend on both major parties' changing some of their ways:

> For its part, Europe must lead in old and new ways. It must continue to reduce its own emissions and press the United States for domestic action while also finding for the first time the will to mobilize even larger international climate funds. Europe also must come to terms with the unfortunate truth that U.S. leadership—even in the age of Obama—is far from assured and that Europe must be prepared to continue leading alone. But the greatest responsibility lies with the United States. To whom much is given, much is expected.

Five years later that assessment still seemed sound and indeed right on target.[37]

5
BRICs, BASICs, and Beyond

The acronym BRIC, referring to Brazil, Russia, India, and China, evidently was coined by a financier in 2001 and soon came into wide usage among bankers, financial journalists, and staffers with the International Monetary Fund and World Bank. The perceived commonality was that all four countries were large and making a "market transition"—that is to say, becoming important free-market economies. For related reasons, they came to be a semi-formal diplomatic bloc, including in the international climate negotiations.[1] By virtue of their strong economic growth, they were becoming significant emitters of greenhouse gases, were coming under pressure from the United States and like-minded countries to join in setting carbon limits, and had a common interest in resisting that pressure. In the past decade, there was a regrouping under the heading of BASIC—Brazil, South Africa, India, and China—with Mexico playing a significant role too.

In truth, the big developing countries that are approaching advanced industrial status have never had much in common, in terms of their emissions and energy profiles. Some, like

Russia—and soon, Iran—are huge producers and consumers of oil and natural gas, with relatively high per capita emissions. Others, like Brazil and Indonesia, have vast forest resources, which makes for some particular problems but also opens special opportunities. China and India, which produce and consume coal on a gigantic scale, are in a class altogether by themselves.

The most important single fact about the BRICs, BASICs, or whatever you want to call them, is that their emissions have grown in the past twenty-five years at a speed and to an extent that was not foreseen when the Framework Convention was negotiated. In 1989, 1990, and 1991, the general expectation was that the developing countries might come to account for perhaps half of world emissions in the next two decades.[2] But by the end of the first Kyoto commitment period, in 2012, China and India alone accounted for about half of global emissions—China, more than 40 percent.

That put China squarely in the sights of those targeting the big developing countries for emissions commitments. India would come next, to be followed by the others, one by one.

If, in taking the position from 1997 on that the United States would accept carbon limits only if China did so as well, it was implicitly a US objective to maneuver China in the direction of accepting limits, that strategy in the short and medium term was a complete failure. In the next fifteen years, China would install new coal generating capacity equivalent to England's entire electric power system every few years. Between 2005 and 2009 it added the equivalent of the entire US coal fleet, which at that time was still producing about half of America's electricity. Between 2010 and 2013 it added half that amount again, burning by the end about four times as much coal as the

United States and almost seven times as much as the European Union.[3] As a result, China surpassed the United States as the world's top carbon emitter around 2006 and, a decade later, was generating about twice the US emissions—roughly ten billion tons of CO_2 versus five billion.[4]

During this same period, automotive traffic in China grew cancerously as well. At the turn of the twenty-first century, south of the Yellow River, one could ride many hundreds of miles on a train without seeing a single motor vehicle. A mere decade later, the country's big cities had been completely transformed by superhighway systems. The combination of fumes from internal combustion engines, emissions from the country's dirty steel industry, and soaring coal pollution made for a full-fledged and highly publicized public health crisis. Even before the industrial take-off of the 1990s, some public health authorities were estimating that a million Chinese were dying each year from air pollution. Today, that would be a very conservative estimate.

That the Chinese regime is facing a public health emergency has been no secret among foreign visitors, China's policy-making elites, or the regime itself. At the grassroots, environmental issues are probably the most common cause of spontaneous protests, making plain that this is not only a moral threat but a political one, too. The regime has every good reason to sharply reduce dependence on fossil fuels, which are at the root of both its air pollution disaster and its skyrocketing greenhouse gas emissions. But its quandary is profound, because quick-and-easy alternatives to coal and oil are not easily found. Switching coal generation to natural gas, the main factor behind the decline in US emissions starting around 2006, does not appear to be an option for China even in the medium term and perhaps not even in the long run. Though some see big potential in its frackable gas reserves, that is a minority opinion.[5] The general consensus is that the reserves are deeper and harder to get at than in the United States and that their exploitation will put even greater demands on water resources—itself a major problem in China and one that appears to be seriously aggravated already by global warming.[6] The government has made a major effort to promote production of synthetic natural gas, to reduce urban air pollution and carbon pollution, as well as the country's import dependence. But that appears to be a major misstep, at least in terms of climate change: "If synthetic natural gas is used to generate electricity, its life-cycle greenhouse gas emissions are about 36–82 percent higher than pulverized coal-fired power," scholars have concluded. And "if used to drive vehicles synthetic natural gas has emissions twice as large as those from gasoline vehicles."[7]

Consumption of natural gas did increase more than twice as fast as consumption of coal and oil in the years 2005–11, but because the increase was from a much smaller starting quantity, the total amount of additional gas used still was outweighed by the additional coal and oil. Essentially the same went for zero-carbon sources of energy, some of which grew even faster than gas. China deployed wind and solar at rates far exceeding US rates from 2005 to 2011—China's six-year increases were 82 percent and 85 percent, respectively—but again, the growth was from relatively small bases. Construction of hydroelectric dams and nuclear reactors also proceeded apace, but the additional electrical capacity was far from enough to compensate for coal's footprint.[8]

Even fifteen years ago a visitor to Beijing would have found plenty of people in policy-making circles who were concerned about climate change, well informed on the subject, and seriously thinking about what China might do to address the problem.[9] But through the next decade, the Chinese government held fast to the line that only the advanced industrial countries were required to limit greenhouse gas emissions. At Copenhagen in 2009, its representatives reacted allergically to any talk of developing country pledges and promised only to keep improving the country's carbon efficiency (the amount of carbon emitted per unit economic output)—a commitment it took seriously and honored, to be sure. Speaking at the UN Climate Summit, in September 2014, Vice Premier Zhang Gaoli reiterated his country's commitment to improve carbon efficiency 45 percent by 2020, relative to 2005; China already had registered an improvement of 28.5 percent by 2013. Looking ahead to Paris, he put the emphasis squarely on the basic Rio principles and the obligation of the industrial states to meet their respective commitments; developing countries should adopt the combinations of green technologies that best suit them. That seemed to suggest that Zhang was holding fast to the Rio and Kyoto principle that only the advanced industrial countries should be required to take serious action on climate.

A month later, however, in October 2014, China and the United States announced a surprise bilateral agreement, committing China to have its emissions peak no later than 2030 and the United States to cut its emissions by 26–28 percent in 2025 relative to 2005. The Chinese indicated that they would try to have emissions peak before 2030 and see to it that the country's nonfossil energy share would grow to 20 percent by 2030. China's adoption of a gas peaking commitment represented a fundamental shift in its diplomatic position and reflected a sharper focus on climate policy at the top level of the Chinese government. For the first time, a president could talk about climate policy without notes, pointed out Robert Orr, who organized the UN Climate Summit for the secretary-general.

China's number two, the premier, had primary responsibility for climate policy, said Orr, and China's top climate negotiator had been in that position for almost a decade and was well known to his foreign counterparts.[10]

The bilateral agreement, negotiated in secrecy, owed a great deal to aggressive and imaginative personal diplomacy on the part of the Americans. Secretary of State John Kerry entertained a top-level Chinese counterpart at a famous seafood restaurant in Boston, taking the opportunity to tell him about how only decades earlier it would not have been possible to eat fish from the city's highly polluted harbor. John Podesta, a top counselor to Obama with special responsibility for climate policy, travelled to China to negotiate details. Obama himself met with China's number-one-ranked vice premier Zhang Gaoli during the UN Climate Summit to seal the deal.[11]

Podesta, who soon afterward would leave the White House to serve as chairman of Hillary Clinton's 2016 presidential campaign, hailed the agreement as a breakthrough; he said that China's clean energy commitment was like adding a whole new China on top of what they already are.[12] That assessment, though perhaps slightly hyperbolic, has been generally accepted as valid, and rightly so. Diplomatically, China's accepting *any* responsibility for limiting its emissions was indeed a breakthrough. But just as indisputably, China's peaking pledge fell far short of what would be needed to keep global emissions consistent with the 2°C goal. The numbers admittedly cannot be crunched with precision. As a developing country, China has not been required to provide the climate secretariat in Bonn with annual statements of its emissions; western estimates—by the Netherlands Environmental Assessment Agency, the US Department of Energy's Energy Information Administration, the International Energy Agency in Paris, the Lawrence Berkeley National Laboratory, and MIT, among others—vary by quite wide margins. The Dutch estimates, which usually are at the higher end, put China's 2013 emissions at 10.3 billion tons CO_2, whereas the IEA and Lawrence Berkeley have them at about 8.5, a difference of more than 1.5 billion tons, an amount greater than Japan's 2013 total emissions.[13] But crude, back-of-the-envelope calculations based on any of those estimates would put China's 2030 emissions as much as three or four times higher than their 2000 level, an outcome bordering on the catastrophic from a global point of view.[14] (Podesta might just as well have said that the agreement would result, in terms of greenhouse gas emissions, in another two Chinas being added by 2030—and a Republican candidate for president in 2016 might conceivably say exactly that.)

At Lima in 2014, the final Conference of Parties before Paris, China resisted efforts to build a strong international review mechanism into the meeting's decisions, which was not in the spirit of pledge and review as originally proposed. Nevertheless, China issued its pledge—the so-called Intended Nationally Determined Contribution, or INDC—on the early side, in June 2015, and its statement was long, detailed, and convincing. It said right off that "China is among those countries that are most severely affected by the adverse impacts of climate change" but that its diplomacy was driven "also by its sense of responsibility to fully engage in global governance, to forge a community of shared destiny for humankind, and to promote common development of all human beings."

The INDC proceeded to enumerate some of the country's achievements in building out zero-carbon and low-carbon energy: a 90-fold increase in on-grid wind power from 2005; a 400-fold increase in solar; a 2.57-fold increase in hydropower; and a 2.9-fold increase in nuclear energy. By 2030, China would seek to cut its CO_2-to-GDP ratio by 60–65 percent, increase its non-fossil-energy share to 20 percent, and boost forest stock volume by about 4.5 billion cubic meters. Specific measures would include development of regional climate plans, policies to discourage excessive concentration of industry and population, promotion of decentralized electricity and the "smart grid," development of emissions standards for key industries, hiking reliance on natural gas to 10 percent of total energy, and bringing installed generating capacity of wind to two hundred gigawatts (GW) and solar to one hundred gigawatts. (The combined three hundred GW would be roughly equivalent to one hundred GW nuclear, or roughly a hundred standard nuclear reactor complexes, adjusting for the intermittent character of wind and solar energy.) Not every part of the INDC was so strong; the section on transportation, for example, is rather weak.[15] But on the whole, the Chinese statement makes a credible impression and belies any residual notion one might have that the country is not taking climate change seriously.

"As a responsible developing country, China will stand for the common interests of all humanity and actively engage in international cooperation to build an equitable global climate governance regime that is cooperative and beneficial to all," the INDC reiterates at the end. Having said that, it goes on to list some specific and potentially troublesome things it expects to come out of the upcoming negotiations: undertakings on the part of the developed countries of "ambitious, economy-wide absolute quantified emissions reduction targets by 2030"; provision by them "of finance, technology and capacity building" to the developing countries; to that end, a scaling up of financial assistance to developing countries, starting from the baseline of one hundred billion dollars per year agreed upon at Copenhagen; creation of an "international mechanism" to take care of poor-country capacity building, and the adoption of "rules" to guarantee "transparency" in the delivery of that financial aid.[16]

It was as if China, having seen the United States insist on a reformulation of the diplomatic game's basic rules during the first decade of the twenty-first century, was now assuming the mantle of most important player in the second decade. It was as if China were saying: If you want us to take aggressive action to help head off catastrophic climate change (which we fully intend to do), then you had better get ready to make some other plans as well. Still, China had gone a long way toward ditching its traditional position that, as a developing country, no real action on climate was expected of it. Especially noteworthy was China's pledge, just months before the 2015 Paris conference, that it would donate $3.1 billion to help less developed countries cope with climate change—a pledge that matched the US commitment.[17]

Having used personal diplomacy to real effect with China, the Obama administration tried to do the same with India, courting the newly elected Hindu nationalist prime minister Narendra Modi. Advocates of a stronger Indian climate policy had attended Modi's election with some optimism: As the leader of the Indian state of Gujarat, Modi had promoted the deployment of wind energy and had written a book trumpeting his achievements, *Convenient Action* (an obvious play on Al Gore's *Inconvenient Truth*); despite his free-market, growth-oriented political base, it was widely felt that here was an unusually strong leader who might—like Richard Nixon, Menachem Begin, or Ronald Reagan—surprise his supporters with some kind of bold about-face on traditional positions.[18] So far such hopes have been unrequited. American diplomats glad-handed their Indian counterparts during the UN Climate

Summit in September 2014, and the following year President Obama tried to twist Modi's arm during a state visit to Delhi.[19] But the Indian leadership flatly refused to countenance any talk of setting a limit to the country's emissions growth—to designate a year when emissions would peak—taking essentially the same position China had adopted until just after the 2014 climate summit.

Particularly disconcerting was the language in which Indian policymakers expressed their antagonism to all such talk. The day after the Climate Summit in New York City, India's environment minister Prakash Javadekar told the *New York Times* that while the government was preparing new climate plans, they would call only for a decreased rate of emissions increase; emissions would continue to increase for at least another thirty years. "What cuts?" Javadekar asked rhetorically. "That's for more developed countries. The moral principle of historic responsibility cannot be washed away."

Tellingly, the environment minister in a previous government came under sharp attack in the Indian parliament for having suggested that India should assume carbon limitation responsibilities, and in the press for allegedly having given too much ground in the Cancún follow-up to Copenhagen on the question of whether developing country pledges would be subject to meaningful review. Chandrashekar Dasgupta, a former India climate negotiator and a member of the prime minister's council on climate change, accused negotiator Jairam Ramesh of agreeing to the principle of "international consultation" at Copenhagen—a weak form of review—which the United States then "pocketed" and "asked for more."[20]

In a characteristic statement of the general Indian attitude, Shyam Saran, a former foreign secretary and current chairman of the national security advisory board, wrote: "While the developing world is advised to adopt low-growth strategies, the developed world appears to be intensifying its own energy and fossil fuel-led growth. India can expose this hypocrisy only if it refuses to be complicit."[21] Saran has argued that to the extent the world is demanding that India unlink economic growth from fossil fuels, the country is being asked to do something no other major nation has ever done before; as such, India is deserving of some special consideration.[22]

Saran, despite that general attitude, takes the problem of global warming very seriously and does advocate Indian action on climate. "Global warming is not only exacerbating all the stresses and strains that have been multiplying over the past several decades, but adding a whole new dangerous dimension to the challenge due to the melting of the glaciers across the entire Himalayan range." Accordingly, he calls for an aggressive shift to clean and renewable energy, including nuclear energy, so as to fully align India's national interests with those of the world. But even though the government in fact is pursuing such policies, there is no expectation that renewables or nuclear energy will be able, on their own, to meet the country's fast-growing energy needs in the next decades. Efforts to promote nuclear energy, going back to the 1950s, have been notably ineffective.[23]

Under those circumstances, if India is to lift hundreds of millions out of crushing poverty and give them the benefits of a modern, electrified life, it sees no alternative but to radically boost reliance on coal. In November 2014, the country's electric power minister Piyush Goyal told a meeting in Delhi that the country would double its coal consumption by 2019, implying further drastic increases in the third decade of the century. Nitin Desai, a retired UN diplomat, put it like this: "Even with the most aggressive strategy of nuclear, wind, hydro and solar, coal will still provide 55 percent of [India's] electricity consumption by 2030, which means coal consumption will be two-and-a-half or three times higher than at present. Mining and burning coal imposes huge environmental burdens. It's a double whammy: The more coal we extract, the more forest we lose, and that too will add to global warming."[24]

India's cities are said now to be as much as three times more polluted than China's.[25] Burgeoning automotive traffic as well as coal-fired power is of course an important factor in India as well. If India is going to avoid being "awash in cars," it will need to emphasize mass transit and railroads, points out economist Rakesh Mohan, who led a major transportation study for the Indian government. But rail systems require electric power, so their rapid expansion also implies a two- or threefold increase in coal combustion. Taking all these factors into account—the need for economic growth, air pollution from coal and cars, expansion of mass transport, still more coal combustion—India's projected future appears to be unsustainable in the literal sense of the term. In projecting such self-destructive increases in coal, which follow ineluctably from current trends, "we're certain we'll be wrong, we're bound to be wrong," comments Mohan, now a member of the governing board at the International Monetary Fund.[26]

Taking a long view of India's climate diplomacy, Joyeeta Gupta observed that its "officials have been very reluctant to accept a constructive negotiating position, arguing that the US should take action first."[27] Like China, India did adopt a carbon intensity goal, promising at Copenhagen to shave its CO_2-to-GDP ratio by 20–25 percent by 2020, relative to 2005. In subsequent negotiations it indicated it might be willing to set some kind of conditional emissions target. Yet the current government's attitude seems essentially unchanged from that of its predecessors, despite Obama's considerable blandishments. As India's Copenhagen negotiator Ramesh put it in an interview with Yale e360, "In India there is this sentiment that, 'We haven't caused the problem, so why should we take tough measures to solve it. It's the responsibility of the developed world.' Mr. Modi has also articulated that view, in the sense of, 'if you want us to do something, give us money and technology.'" At Rio in 1992, India's negotiators said pointedly that if every other country's per capita greenhouse gas emissions were at their country's level, there would be no climate problem to worry about.[28]

A peculiarity of India's diplomatic attitude in recent years has been its tendency to align itself mainly with the emerging market economies—the BRICs or BASICs—rather than the larger mass of developing countries. Given that the world's most crushing poverty is concentrated in South Asia and Africa, it might seem logical that India would take sides with its neighbors and the more active African states. Bangladesh has in fact taken offense at India's orientation.[29] India's alignment with the BASICs seems to follow partly from its shared interest in resisting demands for emissions limits from the industrial countries, but it seems that there is also an element of personal snobbery involved. This anyway is what some Indian critics of the policy quietly suggest. Its politicians and diplomatic representatives would prefer to be associated with the leaders of rising, major states rather than with the struggling masses of the world.

India's per capita GDP, at about four thousand dollars a year, is less than quarter that of Russia's or Mexico's, less than a third that of Brazil, and less than half that of China.[30] It is barely a third of South Africa's and barely a quarter of Turkey's; among the BASICs (in a broad sense of the term), only Indonesia is almost as poor. By the same token, India's per capita greenhouse gas emissions, at 1.8 metric tons per person per year, also are much lower than those in any of the other major rap-idly developing countries. Only Brazil, with an electricity sector predominantly based on hydropower and an automotive sector in which vehicles are often fueled with cane ethanol, comes close to having as low per capita emissions, despite its much greater productivity and wealth. So it is easy to see why Indians feel that the rest of the world has no business asking them to rein in their emissions until their country's per capita income and emissions are much higher.

When it came time to make climate pledges to the world during the run-up to the 2015 Paris conference, India was one of the last countries and the very last major nation to issue its document of intent. Although its INDC said that 40 percent of the country's electricity would come from renewables by 2030, it also clearly implied that coal would still account for nearly three-fifths of the country's power generation. And although the INDC anticipated that the economy's carbon intensity would be cut by a third, the whole emphasis of the pledge document was on growth. Fully one half of what will be the Indian economy in 2030 "has yet to be built," the pledge said.[31]

In August 2014, the Asian Development Bank issued a report concluding that in a business-as-usual growth scenario, six South Asian countries—India, Bangladesh, Nepal, Sri Lanka, Bhutan, and the Maldives—would pay a significant penalty in cumulative GDP by 2050 if they did not scale back their emis-sions to be consistent with the 2°C warming path.[32] But a month later, when Modi appeared at Madison Square Garden in New York to receive acclamation from the many Indian Americans living in the city and its vicinity, he did not mention sustainable development. A video introduction opened with a quotation from Gandhi: "Be the change you want to see in the world." No images of fossil combustion or climate catastrophe followed.

Instead one saw a picture of Modi with Russian president Vladimir Putin, on the occasion of Modi's assuming the titular presidency of the New Development Bank, originally the BRICS Development Bank, a Putin initiative.[33]

Among the BRICs or BASICs, Russia's national business plan (if you will) is uniquely tied to extraction of oil and gas and, to the extent possible, monopolization of Eurasian fossil fuel supplies. As one of the world's two top producers of oil and of natural gas, Russia would seem to be a natural diplomatic ally of the OPEC countries. Yet Russia has played a generally constructive role in climate diplomacy, partly because of specific material interests, but also because it has sought whenever possible to be seen as a good global citizen.[34] During the first Kyoto commitment period (1997–2012), Russia reaped billions of dollars by selling emissions permits in the international market because of the generous emissions targets Kyoto gave it and other former Soviet states. It was Russia's formal ratification of the Kyoto Protocol, in 2005, which brought the protocol into legal force. (In bilateral talks with Europe, Russia was able to trade its support for admission to the World Trade Organization.) Since then, it has continued to refrain from rocking the boat in climate negotiations, leaving the dirty work of defending fossil interests to other OPEC states.

Still, Russia is the world's fourth largest emitter of greenhouse gases, after China, the United States, and the European Union, which people tend to forget because it keeps such a low profile and doesn't make waves unnecessarily. Its emissions are greater than Japan's or Germany's, and so it needs to be more a focus of global carbon-cutting efforts than it tends to be.[35]

As the Paris climate conference approached, Russia was taking an ambiguous and perhaps overly clever position, as some of its influential oligarchs were suggesting that the likely Paris outcome would not be adequate and not in Russia's national interest. Russia's pre-Paris pledge was to cut its emissions 25–30 percent by 2030 vis-à-vis 2005, which superficially looked very similar to the pledges made by the United States and its JUS-CANZ allies at the United Nations (see chapter 6). But in fact, according to the Bonn secretariat's statistics, Russia's current emissions, as of 2013, were already half what they had been in 2005; so 2030 emissions that are 25–30 percent lower in 2030 than in 2005 would actually be much higher than current emissions. In early October 2015, a Russian aluminum oligarch told the *Financial Times* of London that the Paris agreement could carry competitive risks for countries like Russia and yet offer little in the way of emissions reduction and no enforcement. His position seemed to be similar to that taken by some of the US Republican presidential candidates at that time: that it would be foolish for the United States to disadvantage itself internationally just for the sake of getting an agreement that would be so weak.[36]

In its general diplomacy, Russia has sought to rally the BRICs and BASICs to its side, initiating the creation of the Shanghai Development Corporation and the BRIC Development Bank, which it has cast as alternatives to the World Bank and International Monetary Fund for rapidly developing nations. Those organizations have had a distinctly "extractionist" cast, in the considered opinion of Diego Azzi, a climate adviser to Brazil's leading labor organization, the Central Unica dos Trabalhadores.[37] That is, they emphasize the common interest of countries like Russia, Mexico, and Brazil in oil exploitation and of China, India, and South Africa in coal.

Brazil, despite its hosting of the seminal Rio Earth Summit of 1992, was initially quite suspicious of the emerging climate framework, fearful that it would become an instrument for constraining Amazon development. But it came to see—and so did Indonesia—that its possession of giant CO_2-devouring rain forests gave it real diplomatic leverage. In the following decade, both countries were able to exact foreign aid to limit deforestation, and both had considerable success in those efforts, to the whole world's advantage.[38] Brazil's pre-Paris pledge was one of the world's most ambitious: It promised to cut emissions 37 percent by 2025 from 2005 levels and to increase the renewables share in its total energy consumption—not just electricity—from 35 percent to 45 percent. Indonesia's pledge, on the other hand, was rather vague, stating hoped-for emission cuts in terms of "business as usual" projections rather than hard numbers. Mexico's also was imprecise, though Mexico previously had played a constructive role as host of the 2010 Cancún climate conference and had adopted a national climate action plan in 2013.[39] Diplomatically, Brazil, Mexico, and especially South Africa have played consistently positive and often quite important roles in the years since Copenhagen. It was at COP 17 in Durban, two years after Copenhagen, where the world adopted the action plan calling for adoption of a major new, legally binding agreement in Paris in 2015.

Each and every one of the BRICs or BASICs has something to contribute to the world to the extent that each adopts a more sustainable, climate-friendly growth path. But it also is important to recognize their limits. The combined greenhouse gas emissions of Brazil, India, Indonesia, Mexico, Russia, and South Africa today are not even half of China's. What is more, if India were to remain on a path involving doubled coal combustion by 2020 and tripled by 2030, it also will have surpassed the combined emissions of the others. And so, as India's former foreign minister observed in a paper with Bruce Jones, "If India chooses to grow through the same carbon-intensive pathway that has characterized every other major country's growth, there will be no credible prospect for maintaining progress on global carbon reduction. On 'business as usual' projections, India would add another EU to the world's carbon emissions budget within a few decades. On the other hand, denying India the right to grow and confining hundreds of millions to continued poverty is an untenable proposition."[40]

There we have it, in a nutshell: The growth trajectories of countries like China and India are incompatible with the 2°C goal, and yet without following those trajectories, it is hard to see how those countries can climb out of poverty and attain the same standards of living taken for granted among the advanced industrial countries.

6

Sentimental Attachments, Existential Threats

The European Union, the United States, China, and India are the major powers that have dominated the geopolitics of climate diplomacy, but they are not the only powers. Other loose alliances or negotiating blocs also have played significant roles, and so have interest groups and organizations that keep a close eye on the negotiations from the sidelines, among them the environmental organizations that have a global footprint, UN organizations, and—most recently—the Vatican. Some such agglomerations and organizations, starting with the most significant of them, the Group of 77 (G-77), have a distinctly archaic ring.

An observer dropping in on a Conference of Parties for the first time, whether in Copenhagen or Lima, Bali or Durban, may have been startled to find just how prominent a presence the Group of 77 is. For old-timers, the name will likely bring to mind associations with the nonaligned movement of the early Cold War years, with Jawaharlal Nehru and Sukarno, or with the Arab oil boycotts of the 1970s—the 1973 boycott was the first time the OPEC countries managed to sell themselves to developing countries as heroes of the developing world, standing up to the privileged, erstwhile colonial powers.

Founded in 1964, the Group of 77 now includes well over 130 states, though it has retained its original name because of its resonances. Basically it comprises just about all of the world's states that are not members of the other major transnational organizations—the Organisation of Economic Co-operation and Development (OECD), the European Union, or the CIS (the post-Soviet Confederation of Independent States). Thus, for example, Mexico and South Korea dropped out of the Group of 77 when they joined the OECD.

Generally, special circumstances account for the few countries that are members of neither the Group of 77 nor the OECD, European Union, or CIS. Yugoslavia had been an active leader in the nonaligned movement and the Group of 77, but following its civil wars and dissolution in the 1990s, only Bosnia continued as a member of the group. South Vietnam had been a member, but following its defeat and collapse in 1975, the unified Vietnam did not stay in the group. The Pacific island countries Palau and Tuvalu have preferred to affiliate themselves with the Alliance of Small Island States.

The Group of 77 describes itself as "the largest intergovernmental organization of developing countries in the United Nations, which provides the means for the countries of the South to articulate and promote their collective economic interests and enhance their joint negotiating capacity on all major international economic issues within the United Nations system and promote South-South cooperation for development."[1] Pursuant to those lofty goals, its members meet regularly, have some standing staff support, and regularly issue "declarations," "recommendations," and "conclusions" that generally fall on deaf ears. Climate diplomacy represents one of the few areas, arguably, in which the Group of 77 has had some real impact. Diplomatic representatives of the industrial countries ignore the group at their peril, as they first learned when some thirty thousand people assembled in Rio in 1992, for the Earth Summit.

The Group of 77's shared attitude, as one student of the summit put it, was simply this: "There would be no First World today, with a minority of the Earth's population enjoying a comfortable lifestyle, if the majority of the people in what used to be called the Third World had not contributed the incredibly cheap rubber, cotton, minerals, oil, and labor (including slave labor), which enabled the Europeans and Americans to industrialize, produce cheap factory goods, and sell them back to the areas of their empires that have provided the raw materials."[2] To an extent, the very cause of environmentalism has been a casualty of the global South-North conflict. This is because, "basically, the South [was] telling the North that if the latter wanted to indulge in this environmental concern, and expects the South to follow suit, then the North . . . must be willing to pay for the South's participation in this clean-up activity."[3]

From the start, the G-77 countries tended to consider climate change a diplomatic opportunity to exact more financial and technical aid from the industrial world. José Goldemberg, an eminent Brazilian physicist and environmental scientist who was an important player at Rio, got the impression that they even nursed a "fantasy" of turning the Earth Summit into a development summit.[4] Joyeeta Gupta, who is highly sympathetic to the concerns of the Third World, had this to say about the Group of 77 at Rio: "Although many of these countries had geographical, demographic, technological and financial characteristics that made them particularly vulnerable to climate change, preparing for such 'future, abstract' impacts was not prioritized." The whole issue seemed to them imposed from without. Domestically, "there was very little social debate or context-related scientific information, and [so] countries tended to make issue linkages to other issues" like poverty alleviation, development, and food security.[5]

The Group of 77 did play a constructive role at Rio in pushing for a global emissions stabilization target, over the vehement opposition of the United States. At Kyoto, again in alliance with Europe and with lobbyists representing nongovernmental organizations, the Group of 77 got the industrial countries to agree to mandatory emissions cuts. Nevertheless, through the 1990s and right down to the Copenhagen conference in 2009, the representatives of the developing countries continued to leave the impression that they cared more about using climate change as a lever than about the issue as such. Not only their diplomatic adversaries but even their allies in nongovernmental communities worried over this. In early negotiations they rarely seemed able to develop positions that went "far beyond the lowest common denominator," said Gupta; they remained highly susceptible to divide-and-rule tactics and typically "end[ed] up feeling cheated by the final results."[6]

The extent that the G-77 countries have allowed a sentimental sense of solidarity to trump objective self-interest is nowhere more apparent than in their relations with OPEC. Though most of them stand to lose from higher oil prices and from oil-induced global warming, they have tended to side with OPEC diplomatic positions and have allowed OPEC to have excessive influence on the diplomatic agenda. In particular, leader Saudi Arabia found considerable support in the Group of 77 for its argument that countries in the global South should be compensated by rich countries for any costs incurred in addressing climate change. That led, according to an authority on the subject, to "complex, time-consuming and otherwise unnecessary negotiations around the issue of 'the adverse effects of response measures.'"[7] That scholar called the OPEC countries "the worst of friends"; another characterized Saudi Arabia's habitual conduct at COP meetings as "striving for no."[8]

The oil producers have consistently opposed every meaningful action in the climate talks and have seen to it that the actions most effective in principle—a globally harmonized carbon tax, for example, or an end to fossil fuel consumption and production subsidies—never get anywhere near the negotiating agenda. At one Conference of Parties (COP 9), the Saudi Arabian delegates were so unashamed of their designation by an environmental group as "fossil of the day" that they posed for a group photograph in front of the fossil-of-the-day booth.[9] A member of the Saudi delegation once told a staffer for the environmental group that the Saudis considered their receipt of the fossil awards a matter of national pride and strove to get as many as possible.

Over decades of talks, the Saudi delegation came to be seen as not merely very determined but as highly skilled, stated Joanna Depledge, a former staff member of the UN climate secretariat. Its de facto head, Mohamed Al Sabban, was "renowned for his ability to spot opportunities to push his country's agenda forward."[10]

At the opposite extreme from OPEC, another developing country subgroup, the small island states, AOSIS, has exerted a largely positive influence on negotiations. In the preparatory talks for the 1992 Earth Summit the island states lobbied hard for commitments going beyond mere greenhouse gas stabilization.[11] At the first Conference of Parties in Berlin (COP 1), the AOSIS countries played a key role in breaking a logjam, getting diplomats to agree on the "mandate" that led to Kyoto. The Berlin conference

(COP 1) was "rescued ... by some of the least powerful states in the United Nations, aided by hardworking coalitions of NGOs from North and South. The AOSIS countries, ... joined by the rest of the developing world, ... were finally able to persuade the OECD nations to agree on [a mandate] that established a process and a timetable for negotiating further actions."[12]

Like dying patients who get a lot of attention even though there may be nothing anybody can do to save them, AOSIS representatives are listened to when they speak up at the annual climate meetings.[13] And when citizens of their countries take direct action—using traditional canoes to obstruct Australian cargo ships sailing to China with coal, for example—they get attention in the world press.[14]

The Group of 77 is not the only significant negotiating bloc that may be bound by sentimental attachments more than anything else. Just as anomalous, in its own way, is the group known among students of climate diplomacy as JUSCANZ, for Japan, the United States, Canada, Australia, and New Zealand—or sometimes that group plus Norway and Switzerland. At first glance, anyway, its members seem to have not much in common except wealth and a certain comfort with the English language. Whatever its rationale as a negotiating bloc, it has had considerable influence on climate negotiations, both for better and for worse, and individual members have had considerable influence over one another, also for better or worse.

It was Japan, in the Rio negotiations, that first came up with the idea of pledge and review, which was considered one of the weaker ideas on the table at the time.[15] It was rejected at Rio, despite US support, and again at Kyoto, only to be revived at Copenhagen in 2009 and made now, almost entirely at the behest of the United States, the basis of the long-term climate agenda.

Japan also is credited, though, with actively mediating as host at the 1997 Kyoto conference, helping parties agree that the industrial countries should seek reduction of their emissions by 5.2 percent by 2008–12. That is, it nudged parties toward that compromise despite its own ambivalence about the whole idea of mandatory emissions cuts.[16] Generally Japan has played a consistently moderate and moderating role in climate negotiations.

Canada's position has been much more erratic. Having agreed at Kyoto in 1997 to reduce its emissions 6 percent by 2012, the Canadians instead saw them rise by nearly 25 percent in the next ten years. Under the influence of its neighbor to the south, Canada elected an archconservative government in 2006, which staked the country's destiny wholeheartedly on all-out development of the Alberta oil sands, probably the most carbon-intense of all fossil fuel sources. In 2011, after the United States derailed the Kyoto process at Copenhagen (chapter 9), Canada formally repudiated the Kyoto Protocol. Its initial pledge at Copenhagen, and again its more formal national pledge in 2015 (its INDC), both said that it would pretty much do whatever the United States would do. (Since the United States had promised to cut its emissions by 28 percent by 2025, Canada said that it would cut its emissions 30 percent by 2030.) It was a strange attitude to adopt for a country that often has styled itself, like the Scandinavians, as a moral world leader.

Much of the debate in North America over Canada's energy and climate policies came to focus in the years 2013–15 on the proposed Keystone Pipeline, which could carry Alberta crude down to the Gulf of Mexico. By now, Canada was exporting two million barrels of crude oil to the United States per day, accounting for about two-fifths of US oil imports.[17] US environmental groups unanimously opposed Keystone, and members occasionally chained themselves to the White House fences to demonstrate their antagonism to it. Some of their leaders linked the project directly to Canada's position on climate change, but more commonly they argued against the project on the basis of narrow, local environmental concerns. Though the project was under continuous State Department review, its connection to Canada's climate policies went unmentioned, presumably because the United States was in a poor position to criticize Canada for having followed its lead out of Kyoto. At the uppermost level of US government, there probably was a reluctance to frame *any* issue in terms of climate change unless absolutely necessary.

Former New York City mayor Michael Bloomberg, a leader in global efforts to organize cities to address climate change and the UN secretary-general's special envoy for cities and climate, was a lone voice in calling for the United States and Canada to take the focus off Keystone and jointly address the larger issues of global warming. "The Keystone XL pipeline has become a perfect symbol of Washington's dysfunction. Democrats exaggerate its environmental impact while Republicans exaggerate its economic benefits," he wrote. Since the Canadian government has been pressing for its approval by the United States, that "gives the White House enormous leverage, which it should use to negotiate a broader, climate-friendly deal."[18] Bloomberg must have been satisfied with the ouster in October 2015 of the conservative Canadian government that had firmly commited for nine years to all-out development of fossil fuels and its replacement by a government much more open to constructive climate policy and diplomacy. On the eve of the December 2015 Paris climate conference, Obama nixed the pipeline project.

Despite their many obvious defects, Canada's climate policies have resembled those of the United States in positive as well as negative ways. Individual provinces, notably Ontario and British Columbia, have adopted ambitious climate action plans that resemble those formulated by California and some of the northeastern states. British Columbia was the North American pioneer in introducing an economywide carbon tax, and Saskatchewan was first in the world to bring a carbon-capturing power plant into operation, the Boundary Dam facility.[19] Both Ontario and Quebec have indicated they will participate in the carbon trading market being set up by California, which means that at least three quarters of Canadians will live in provinces where a price has been put on carbon.[20]

Australia is the other really significant Anglophone emitter among the JUSCANZ countries: Australia's and Canada's emissions each amounted to somewhat more than a half billion metric tons per year in 1990; together, they were roughly the equivalent of Japan's. Since then, Australia's domestic emissions have remained flat, unlike Canada's, but the country's coal exports to China and India have soared, making the future of coal the most important factor in its wildly oscillating climate policies. After devastating wildfires in 2009 and in the throes of ongoing acute water shortages, Australia enacted a national carbon tax in 2012, only to repeal it in 2014, after a conservative government took power the year before. (According to a legislative survey, it was the only one of sixty-six countries studied to reverse a climate law that year; of 487 climate laws that have been enacted worldwide, only Australia and Canada have repealed laws.)[21] Nevertheless, the new government led by Tony Abbott stuck with the country's pledge to cut emissions 5 percent by 2020 in the second Kyoto commitment period, even though the second Kyoto phase was barely alive, diplomatically. Australia met and indeed somewhat exceeded its first-phase Kyoto target.[22]

A notable aspect of the climate policy debate in Australia is its vitriol. Abbott was widely reported to have called climate science "crap" and coal "good for humanity." In his victorious campaign, he promised "in blood" to "axe the [carbon] tax." The man Abbott ousted as leader of the conservatives called the prime minister's climate plan, which provides grants to companies and organizations that voluntarily cut their emissions, "bullshit" and "a recipe for fiscal recklessness on a grand scale." The climate minister in the previous government, Penny Wong, said that Abbott would go down in history as "one of the most short-sighted, selfish and small people ever to occupy the office of prime minister."[23] At the end of the summer of 2015, with global concern about climate change rising perceptibly and the world's attention increasingly focused on the upcoming Paris conference, Abbott was ousted in a party coup and replaced by somebody more positively inclined on climate policy. Together, the Australian and Canadian reversals seemed a striking demonstration of how somewhat subtle global trends can influence domestic politics in one country or another.

Considered in general terms, the JUSCANZ grouping would seem to be in its own way just as vestigial as, say, the Group of 77, and yet it still seems to have a certain mysterious coherence. As countries were maneuvering and adopting positions in preparation for the December 2015 Paris climate conference, Japan, the United States, Canada, and Australia all pledged to cut emissions in the range of 25–30 percent by 2025 or 2030 relative to 2006. Norway, on the other hand, aligned itself with the European Union's 40 percent by 2030 pledge, while Switzerland promised cuts of 35 percent by 2025 and 50 percent by 2030.

Julie Bishop, the foreign minister in Abbott's government, once said that she did not appreciate until taking office just how much "the climate issue matters for Australia's international standing." When she and other representatives of countries like Australia and Canada show up for major meetings like Paris, they must reckon on being given a hard time not only by their diplomatic counterparts but by the many nongovernmental organizations closely monitoring the proceedings from the sidelines. The global environmental organizations have been a major presence from the beginning, leading up to Rio and at the Earth Summit itself, where many participants found the general atmosphere to be more like a Latin fiesta than what one would expect at a high-stakes negotiation. The environmental NGOs pushed for creation of the Intergovernmental Panel on Climate Change in the late 1980s, the initiation of the talks that led to Rio, and the Berlin Mandate that set the stage for Kyoto, notes Alden Meyer of the Union of Concerned Scientists, an NGO leader who has been at just about every Conference of Parties since COP 1 in Berlin.[24]

At Kyoto, the NGOs—formally organized under the umbrella of the Climate Action Network—had a distinct and possibly decisive impact on the outcome of the negotiation. First, they did much to rally the Europeans and G-77 states to press for a global stabilization target and binding greenhouse gas reductions commitments from the industrial nations. Second, they were highly critical of proposed trading systems, the idea of taking carbons sinks into consideration in calculating emissions levels, and especially the generous allowances being given the states of the former Soviet bloc, which they dubbed "hot air." It was under these circumstances that outside experts like Michael Oppenheimer of the Environmental Defense Fund (EDF) and Dan Lashof of the Natural Resources Defense Council (NRDC), the two US environmental organizations best known for mustering top scientific and legal talent, came to be seen as players just as important as many a country's formal representatives.[25] A key turning point in the negotiation was when Vice President Al Gore, who was very close to the environmentalist community, flew into Kyoto and asked the US delegation to be more flexible and accommodating on the question of mandatory emissions cuts, which the NGOs were pressing for. Though the United States prevailed on emissions trading, carbon sinks, and "hot air" in the compromise that emerged, the nongovernmental organizations would have a constructive influence in the coming years in making implementation of those concepts more rigorous.

During the eight years of George W. Bush's presidency, following its repudiation of the Kyoto Protocol, the Climate Action Network (CAN) consistently urged all other industrial countries to stick with the program and disregard US nonparticipation. "There was no one in the American environmental community who wanted to support Bush's position, and back off on the issue of mandatory emissions cuts," says Meyer.[26] Yet since Copenhagen, when negotiations suddenly took a quite different and unexpected direction, the NGO community sometimes has seemed nonplussed. At the end of 2009, Greenpeace labeled Copenhagen a "crime scene," but a year later at the Cancún COP, which did little more than ratify Copenhagen, Greenpeace said that "negotiators have resuscitated the UN talks and put them on the road to recovery."[27]

During the decade before Copenhagen and especially during the years immediately preceding the conference, the number of NGOs showing up at the annual climate conferences increased dramatically. Tactics changed, too, with many NGO members coming in order to stage public demonstrations, engage in guerrilla theater, and even perform acts of civil disobedience. To a great extent those new recruits were drawn from the ranks of people who had been demonstrating against globalization and its perceived main agents, going back to the Seattle WTO protests of 1999. According to one close student of the subject, by the time of Copenhagen, the number of NGOs that were primarily concerned about climate change—as opposed to using the issue of global warming as a lever to address other issues—had actually become a minority within the Climate Action Network. Within CAN, an "equity caucus" attuned mainly to issues of climate justice dubbed itself the Rising Tide Network.[28]

Since Copenhagen, the Climate Action Network's influence on events appears to have waned some, as divisions in its ranks have deepened, its annual presence at the COP meetings has become increasingly ritualistic, its leadership has aged, and—arguably—the older leadership has come to see itself as part of the diplomatic establishment.

Naomi Klein, in *This Changes Everything*, delivered a scathing critique of some NGOs for accepting funds from fossil energy companies, and of some NGO leaders like the EDF's Fred Krupp for accepting huge salaries.[29] She opened her chapter on what she called "the disastrous merger of big business and big green" with a telling story about how the Nature Conservancy, "the richest environmental organization in the world," allegedly got into drilling for oil and gas in a Texas nature preserve. Whatever the ultimate outcome, that episode captures "the general failure of the environmental movement to confront and challenge the economic forces driving up greenhouse gas emissions," she said.[30]

Klein may have a point. If negotiations continue to produce results that fall far short of what is needed to head off catastrophic climate change, some NGO members may have to look more critically at the cozy relations that have developed between their leaders and political establishments. At the grass-roots, more radical tactics may indeed be called for, including civil disobedience. Yet at the same time, it pays to be sensitive to the limits of what grassroots activism is likely to deliver over the long haul. Among activists and activist leaders, there has been a lot of talk comparing today's climate movements with the antinuclear weapons and antiwar movements of the 1960s, 1970s, and 1980s. Without a doubt, those movements had a huge impact on US policy in the Vietnam and Reagan years, leading to dramatic reversals of direction. But once those reversals had taken place and the US government had struck radically new courses, the hundreds of thousands of Americans and Europeans who had been turning out for demonstrations abruptly disappeared from the scene. After Reagan's 180-degree turn on nuclear weapons, furthering arms control would be the work of professional diplomats, and no more would be heard of the US Freeze Movement or the European Nuclear Disarmament campaign.[31]

In a famous letter addressing the issue of historical determinism and human freedom, Friedrich Engels said, "We make our history ourselves, but in the first place under very definite assumptions and conditions."[32] Especially in diplomatic history, the role of personality—animal magnetism, charisma and charm, intellect, dogged determination—is not to be underestimated.

Among the personalities who can have a significant impact on climate negotiations are the individuals delegated by the United Nations itself to manage the talks: Those can include the person delegated by the host country of a COP to chair the daily meetings; other major figures in the host country government; the men and women participating in the work of special task forces or preparatory committees; and UN officials like Robert Orr and Selwyn C. Hart, the able deputies to Secretary-General Ban Ki-moon in all matters pertaining to global warming.

The role of the UN Framework Convention's secretariat in Bonn is "that of a traditional civil service at the international level," observes Richard Kinley, who has worked in the secretariat at a top level from the beginning. "The secretariat cannot be seen as supporting any one country's position." However, he adds with some emphasis, the secretariat is "*not* neutral on the question of whether a good climate treaty is desirable."[33] The situation of the IPCC's much smaller secretariat in Geneva is closely analogous, observes Gaetano Leone, who until recently was second in command there. Its member governments are in effect the IPCC, and "they determine what goes into the summaries for policymakers," proceeding by consensus and unanimity, the same way the conferences of parties work.[34]

Around the time of the Copenhagen conference in 2009, some embarrassing disclosures of errors had crept into IPCC reports, which in turn led to a tightening of procedures and efforts to qualify findings always in terms of explicitly stated probabilities. Unfortunately, the net result has been to make the IPCC reports less readable, and the organization still stands accused of operating less transparently than it might. Nonetheless, it seems to have survived the big storms, and its conclusions are increasingly accepted without much controversy.[35] Perhaps its bureaucratic opaqueness and obscurity, in its own way, is a good thing.

Commenting on the role of individuals (as opposed to formal government organizations) in diplomacy—in this case in the creation of the International Criminal Court—UN rights chief Prince Zeid Ra'ad Zeid had this to say: "It's not governments that brought this court about, it really was 60 individuals who decided they wanted the court. I'm not sure if there had been 60 other individuals representing the same governments you would have had it."[36] In climate diplomacy there have been many instances in the past where unusual personalities have taken negotiations in unexpected directions, and undoubtedly this will be the case in the future as well.

Among the individual personalities whose presence might be felt at Paris, one in particular stood out: Pope Francis. The abrupt emergence on the world stage of Francis is a diplomatic element that was wholly unforeseen and unforeseeable when the agenda was set for Paris in 2009, 2010, and 2011. In 2014, the pope disclosed that a major papal encyclical on humankind and its environment was being prepared, and in the following year he spoke with astonishing transparency about the details of how it was being drafted, how he was having it reviewed and edited, and what his own role in the process would be.[37] It was well known, far in advance of the encyclical's release, that it would draw heavily on earlier church thinking about the relation of capitalism, society, and spirituality, notably the seminal *Rerum Novarum* of 1891. By the time *Laudato Sí* ("On Care for Our Common Home") was issued in June 2015, in fact, the effect was admittedly rather anticlimactic, since it was already pretty well known what its line on climate change would be. Reports and commentaries dwelled little on its actual content, which in truth had nothing original to say about the climate problem as such. What is more, though priests around the world were supposed to deliver sermons about its message and church bells were supposed to toll at assigned times, little of this appears to have taken place. To the extent that it did, little notice was taken.

Frankly speaking, the Catholic Church is itself something of a sentimental attachment. It is no secret that hundreds of millions of professed Catholics go through the motions of fidelity in the most minimal way and that hardly any of them pay serious attention to Vatican pronouncements on subjects like birth control. A millennium has passed since popes were able to force emperors to crawl at their feet—and not much time has elapsed since Stalin asked sarcastically how many divisions the pope had.

Nevertheless, the fact that Pope Francis firmly aligned the Catholic Church with mainstream climate science and enumerated the major ill effects expected from global warming was in itself immensely important. A particularly striking and personal element in the encyclical was the pope's denunciation not only of climate denial but also of a tendency to resignation and indifference, even among those who understand and acknowledge the validity of climate science. Repeatedly, at key junctures in the document, Francis characterized an absence of caring about the world as a kind of sinfulness. He devoted a startling amount of detailed attention to environmental diplomacy and, on the subject of climate, said, "We cannot fail to ask God for a positive outcome to the present discussions, so that future generations will not have to suffer the effects of our ill-advised delays."[38]

One immediate effect of the encyclical, surely, was its inspirational impact in the world of Islam. About a month after the issuance of *Laudato Sí,* Islamic intellectual leaders meeting in Istanbul adopted an Islamic Declaration on Global Climate Change, which denounced humanity's greed for resources, echoing Francis's insistent critique of excessive consumerism, especially on the part of the wealthy. The Islamic declaration called on believers to respect nature's "perfect equilibrium" and recognize a "moral obligation" to conserve.

Even more important here than the church and its doctrine, however, is simply Pope Francis himself. The main reason why his injection of himself into the climate fray could be hugely important is because he, at present, is the only real megapersonality on the world stage. Merkel and Obama are not trivial political personalities by any means. But both carry a good deal of baggage: Merkel's reputation suffered from her handling of the Greek financial crisis, and perhaps somewhat from her uneven reactions to Putin; Obama, in climate circles, has his conduct at Copenhagen to account for and, in Europe, his government's spying on friends. Pope Francis, in contrast, is intact. As the final preparations for Paris were being made, he seemed perhaps the one and only person, despite not being a direct participant in negotiations, who might push them in some unanticipated direction.

III

The Action

The ideal ambassador, wrote the Renaissance diplomat Ottovaniano Maggi in 1596, "should be a trained theologian, should be well versed in Aristotle and Plato, and should be able at a moment's notice to solve the most abstruse problems in correct dialectical form; he should also be expert in mathematics, architecture, music, physics, and civil and canon law. He should speak and write Latin fluently and must also be proficient in Greek, Spanish, French, German and Turkish. While being a trained classical scholar, a historian, a geographer and an expert in military science he must also have a cultural taste for poetry. And above all he must be of excellent family, rich and endowed with a fine physical presence."

—HAROLD NICOLSON, *Diplomacy* (1988 edition)

The secrecy in negotiations to which their contemporaries so often objected is accorded them [the ambassadors] in full measure by the silence of posterity.

—JULES CAMBON, *The Diplomatist* (1931)

7

The Road to Rio

By the late 1980s it was evident that humanity was having an alarming impact on the world's climate. From the 1950s Charles Keeling had been taking direct measurements of carbon dioxide in the atmosphere at a station in Hawaii and had shown that its concentration was steadily rising. Increasingly sophisticated and highly powered computer models affirmed and reaffirmed many times over that the higher levels of greenhouse gases would increase average global temperatures. Not least, ice scientists drilling in Greenland and Antarctica had teased out the history of Earth's temperatures and greenhouse gases going back almost a million years: They found that current levels of CO_2 and methane were without precedent in the history of humankind—and that sometimes in the past, drastic climate changes had occurred astonishingly fast, within decades, not just centuries or millennia. What directly catalyzed and inspired climate negotiations, however, was none of that as such. Rather, it was the ozone diplomacy of the late 1980s, which seemed to show that humankind could take scientific findings, agree promptly on needed remedial actions, and implement an effective, globally coordinated program.

In the 1970s it had been discovered that widely used aerosols containing chlorine, the chlorofluorocarbons (CFCs), were devouring the atmospheric ozone layer that protects humans from one form of ultraviolet radiation (UVB). "A perfume spray in Paris helps to destroy an invisible gas in the atmosphere and thereby contributed to skin cancer deaths and species extinction half a world away and generations to come," wrote Richard Elliot Benedick, chief US negotiator in the ozone negotiations, looking back on them.[1] In 1985, meeting in Vienna, Austria, the twenty top CFC-producing countries agreed on a convention that established a framework for ozone negotiations. Considering those talks in hindsight, from the perspective of the climate negotiations they soon inspired, what first of all is striking is that almost all the roles were reversed. It is as if one went to see a two-act play in which all the villains of the first act turned into the heroes of the second act, and vice versa.

The Europeans, who would generally be the leaders in the first decades of climate diplomacy, would be the main foot draggers. The developing countries, which would turn into such a noisy bunch in climate negotiations, barely showed up. The countries that later would be bundled into JUSCANZ, notably Australia, Canada, and Norway, rather than hanging back, often nudged the ozone talks in constructive directions. Most notably, the United States, headed up at that time by President Ronald Reagan, was the world leader in pushing hard for total elimination of CFCs, on a tight, demanding schedule.

The Montreal Protocol to the Vienna Convention, concluded on September 16, 1987, called for five categories of CFCs to peak in the major producing and consuming countries in 1991–92. By 1994, CFC consumption and production were to drop back to no more than 25 percent of their 1986 level. By 1996, their level was to be zero. The protocol also established a phase-out program for hydrochlorofluorocarbons (HCFCs), which were to be frozen in 2013 and then reduced progressively, starting in 2015. The expectation was that the HCFCs would be a suitable transitional replacement for the CFCs, because their ozone-depleting potential is much lower. Thereafter they in turn would be replaced by hydrofluorocarbons (HFCs), which would be virtually harmless in terms of ozone. Notably, all three gases have very high global warming potentials, as much as ten to eleven thousand times than of carbon dioxide, per molecule. So, the elimination of CFCs may not have had only a modest impact on the warming, given that the HCFCs tended to take their place. In the long run, on the other hand, the effort to discourage use of all these aerosols ought to have some real positive impact on the climate.

The Montreal Protocol took force on January 1, 1989, and, together with the Vienna Convention, came to be ratified by 197 countries. As such, they are said to be the first treaties in UN history to win universal ratification.[2] There ensued several years of follow-up negotiations, pitting nations like the United States that sought more universal participation in CFC and HCFC cutbacks, against countries like China and India, which advanced the notion that richer countries should provide money to help poor countries avoid reliance on the aerosols.[3] UN Environment Programme executive director Mostafa Tolba, and Kenya's president Daniel arap Moi, both eminent third world statesmen and spokespersons, played significant roles.[4] A "grand bargain" was reached in London in June 1990, in which industrial countries promised finance and technological transfer in exchange for complete global elimination of CFCs.[5] In the United States, Congress embraced the amendments to the protocol, as did representatives of nongovernmental organizations and industry worldwide.

At the time the London and Montreal treaties were adopted, Europe accounted for about 45 percent of world CFC production and the United States for about 30 percent.[6] That difference—together with Dupont's development of a chemical alternative to CFCs—may partly account for their respective position in the initial negotiations (though by the end of the 1980s, Europe had joined the United States as an advocate for universal action). Even so, President Reagan's strong personal leadership on the issue appears to have been key: At a June 1987 meeting in Venice, he put the ozone problem at the top of the global environmental agenda. His position "confounded his traditional ideological allies" just as much as his dramatic turnaround on nuclear arms control and disarmament, as Benedick observed.[7] Evidently an important factor was the growing influence in the White House of moderates like Secretary of State George Schulz and Chief of Staff Howard Baker. But Benedick also credited Ed Meese, a bête noire among American liberals at the time, with chairing a committee that cajoled and muscled US industry into getting onboard.[8]

Certainly the discovery of the large and growing ozone hole over Antarctica, just as negotiations were getting organized in 1986, had a big impact on the diplomatic ambience. Nevertheless, anybody surveying the story of how the Vienna and Montreal treaties were negotiated and amended will be struck by the crucial role played by individual human beings, whether political leaders of states like Reagan, international civil servants, or professional diplomats. Benedick himself was important. Another person not to be left out of account was Canada's Maurice Strong. Though he later would come under a cloud because of alleged financial improprieties, Strong had been a pioneer in UN action on the environment and was a living inspiration to his peers at Montreal. He was widely thought of as the main person behind the first UN meeting on the environment, in Stockholm in 1972, and he had served as first head of the UN Environment Programme, which came out of Stockholm. Hence his appointment, five years after Montreal, as the secretary-general of the UN Conference on the Environment and Development at Rio—the 1992 Earth Summit.

Preparations for the climate track at the Earth Summit began during the same years that implementation of the Montreal Protocol was being negotiated and owed much to the ozone negotiations. In 1988, UNEP and the World Meteorological Organization (WMO) jointly established the Intergovernmental Panel on Climate Change, which in the coming decades would make itself the world's main authority on developments in climate science and thinking about climate impacts. Essentially a volunteer effort involving many thousands of scientists around the world, the IPCC would also be seen, in effect, as the expert support body for the ongoing climate negotiations, although they were just being conceived at the time of its establishment. At a Group of 7 meeting held in Toronto in June 1988, a declaration on the environment took note of the Montreal Protocol, called on UNEP and the WMO to establish the IPCC, and welcomed an experts' Conference on the Changing Atmosphere that would convene a week later, also in Toronto.[9] The experts' meeting would produce a call for international negotiations on climate change and conclusion of a framework convention, to govern further negotiations, like the Vienna Convention that had set the stage for Montreal.[10]

The tiny island state of Malta put climate change on the UN General Assembly's agenda, and in December 1990 the assembly authorized establishment of an Intergovernmental Negotiating Committee (INC) to prepare the way for finalizing a Framework Convention on Climate Change (FCCC). A seasoned French diplomat named Jean Ripert was put in charge of the INC, and during the next year and a half, by all accounts, he did a brilliant job. Patiently proceeding by consensus without ever taking a formal vote, but boldly employing the powers and influence of the chairmanship whenever necessary, Ripert got the members of his committee to bridge fundamental national cleavages and arrive at formulations that would closely resemble those actually adopted at Rio. The end result would be another major environmental treaty, of startling vision, which like the Montreal Protocol would come to win universal ratification among the UN's 190-plus members.

The spiritual foundation for Ripert's work, no doubt, was the can-do optimism that had come out of Montreal. When the diplomatic work that led to the protocol was beginning, Benedick recalled, "Knowledgeable observers had long believed that this particular agreement would be impossible to achieve because the issues are so complex and arcane and the initial positions of the negotiating parties so divergent."[11] Even today, a quarter century later, the protocol stands out as "the most successful global environmental agreement in history and a shining example of a current generation taking extraordinary precaution in avoiding environmental impacts to future generations," as a recent scholarly assessment put it.[12]

At the time of its adoption, with the protocol in the bag, people tended toward what in hindsight was an almost myopic optimism: That is, they were inclined to believe that if the tricky technical problem of ozone depletion could be tackled so successfully, why not climate change as well? What they were overlooking, we can now easily see in retrospect, was first of all that the Montreal negotiations had started small and only gradually evolved in the direction of universality.[13] Even more important is the obvious fact that the climate problem is much more all-encompassing than the ozone issue, cutting to the heart of the energy that supplies virtually all our needs and sustains our whole way of life. Thus, though it may have been surprisingly easy at Rio to agree on some fundamental principles and attitudes, it would be much, much harder to reach consensus on how those axioms should be put into practice.

In the next decades, throughout the world but especially in the United States, the fossil fuel industries would launch an all-out assault on the fundamental principles of climate science, using many of the same techniques and indeed much of the same organizational infrastructure that the tobacco industry had used to fight health warnings.[14] Some eminent scientists and public intellectuals would lend their names to the campaign. With the advanced industrial economies in recession in the early 1990s and on the brink of an outright depression in 2008–9, there was wide public support for the notion that one should avoid costly actions to slow and arrest global warming, even if climate change was indeed taking place. And with global competition on the rise from the fast-growing economies of East and South Asia, labor unions in the advanced industrial countries of Europe and North America tended to be sympathetic to the arguments advanced by climate skeptics.

All of that, in a nutshell, accounts for why it would take the world almost twenty-five years to reach at least a tentative consensus on how to realize the fundamental principles of the Framework Convention.

8

Rio and Kyoto

Going into the Rio Earth Summit in June 1992, European negotiators sought to make the advanced industrial countries subject to mandatory greenhouse gas reductions while their US counterparts resisted firm targets. The developing countries and AOSIS generally supported Europe's position, while Canada, Australia, New Zealand, and Japan sided with the United States. Considering those circumstances, the resulting treaty negotiated at Rio, the UN Framework Convention on Climate Change, was surprisingly strong. Legal scholars have called it a "kind of constitution for international action on climate change."[1]

The basic objective of the Framework Convention, as stated in the treaty, is to prevent "dangerous anthropogenic [that is, human-induced] interference in the climate system." Parties to the treaty are to take precautionary measures to "anticipate, prevent or minimize" climate change; lack of full scientific certainty should not be used as an excuse to postpone needed measures. In the effort ahead, a global division of labor is to be based on "common but differentiated responsibilities and respective capabilities." That is to say, the richer countries are to take the lead in adopting countermeasures, giving "full consideration to the developing countries, especially those most vulnerable to the adverse effects of climate change."[2]

The Rio Framework Convention was specific about just what should be done next and how. Parties to the treaty were to submit national greenhouse gas inventories, implement national plans to mitigate climate change, and cooperate in the exchange of relevant technology and scientific research. Each party was required to make regular progress reports to the convention's governing body, the Conference of Parties, or COP, which would meet annually. Taking account of their differentiated responsibilities and respective capabilities, parties were divided into basic categories: The Annex 1 industrial countries, which included both the established capitalist countries and the so-called economies in transition of the former Soviet Union and Eastern Europe; a subset of those countries, Annex 2, which consists basically of the OECD states and would be expected to provide assistance to the developing countries; and everybody else, that is to say, the "non-annex" states.[3] All parties were to report to the convention's central secretariat, soon to be established in Bonn, Germany, where it would make itself the world's authoritative central source of information on how parties were doing in meeting their obligations.

The Rio treaty did not include the specific commitments to cut greenhouse gas emissions that the Europeans had sought. But parties to the convention, including the United States, committed themselves to a process in which such commitments were soon negotiated. The term "framework" implied that stronger and more specific actions would follow, much as the Montreal Protocol governing CFCs came pursuant to the Vienna Convention, as the US ozone negotiator Benedick observed. What is more, said Benedick,

contrary to conventional wisdom, the climate convention was actually a much stronger agreement
than its analogue framework treaty on ozone, the
1985 Vienna Convention. Where the Vienna Convention was limited to cooperation in research and
exchange of information on the ozone layer, and
did not even contain any mention of CFCs, the
Framework Convention on Climate Change embodied commitments to reduce greenhouse gas
emissions, with specific reference to carbon dioxide. Certainly the American coal and oil interests
and anti-environmental ideologues, who had opposed the negotiations every step of the way, clearly
understood what had happened and were furious
with the [George H. W.] Bush administration for its
last-minute concessions.[4]

Especially noteworthy, considering the enormous economic
interests at stake, was the convention's explicit statement that
international action should not be deferred just because full
scientific certainty had yet to be obtained.

President Bush's team would come under some criticism
subsequently for being too beholden to industrial interests and
for having tried to dig in its heels on the issue of greenhouse gas
stabilization.[5] Nevertheless, it is rather startling that universal
agreement was reached so quickly at Rio on fundamental principles that could and would have far-reaching consequences, and
especially that the US leadership was able to go along with the
accord without ruffling feathers at home.[6] The fact of agreement,
besides owing much to the work of Jean Ripert's Intergovernmental Negotiating Committee, benefited from other circumstances as well. During the previous two years, all the OECD

countries except for the United States and Turkey had adopted stabilization targets.[7] And when the global climate talks began, according to climate expert Joyeeta Gupta, international discussion of needed actions did *not* reflect the neoliberal-neoconservative rhetoric surrounding UN discussions of water and energy.[8]

As for the last-minute acquiescence of the US president— and, even more important, his confidence that he would be able to sell the treaty at home—remember that the senior Bush was fresh from a victorious war against Iraq's Saddam Hussein, an enterprise for which he had secured the virtually unanimous endorsement of the United Nations; this was a masterful diplomatic accomplishment by any reasonable reckoning. And that was not all. He and his right-hand man James Baker had coordinated the dramatic diplomatic developments that had brought the Cold War to an end, resulting in the reunification of Germany and the dissolution of the Soviet Bloc, with NATO (North Atlantic Treaty Organization) left intact.[9] If anybody was invulnerable to attacks from America's political right in matters of national security, it was this US president.

Under the senior Bush's successor, President Bill Clinton, US climate policy did not change radically, despite the presence in the White House of the environmental visionary Vice President Al Gore. The United States continued to resist immediate greenhouse gas emissions cuts. But three years after Rio, at the first Conference of Parties in Berlin, the United States found itself outflanked and having to agree to a "mandate" to negotiate reductions in rich-country (Annex 1) emissions. Those negotiations would culminate four years after that in the adoption of the Kyoto Protocol.

The 1997 Conference of Parties at Kyoto was the first major climate meeting to see intense lobbying from environmental organizations and equally intense, if momentary, attention from the world press. Some 265 ecology-minded NGOs were on the scene, among them the most famous on the global stage (such as Friends of the Earth and Greenpeace) and those most influential in US politics (Natural Resources Defense Council, Environmental Defense Fund, and so on). By now, all well-informed and concerned world citizens knew about the issues dividing the United States and Europe, and those represented at Kyoto were keen to influence the proceedings.

Much more than coordinated rich-country emissions cuts was at stake. In exchange for agreeing even to talk about cuts, US negotiators were calling for the developing countries to make corresponding commitments—an idea about which the Europeans, fearful of losing their third world support, were lukewarm at best. In addition, the United States wanted to see account taken of positive changes in land use—modifications of agriculture and forestry leading to more absorption of carbon dioxide from the atmosphere. (Highly forested nations like Russia and Canada liked the idea, whereas skeptics worried about the loss of quantitative rigor likely to attend any such widening of carbon accounting.) Most of all, US negotiators were eager to have Kyoto endorse market-based means of reducing greenhouse gases: basically trading schemes like the one that had been used to such good effect to reduce sulfur emissions in the United States. American negotiators also tended to favor "offsets," credits that one party could get for balancing another party's increased emissions with cuts.

The US efforts at Kyoto to obtain concrete commitments from the developing countries did not go down well with them, their European patrons, oil-producing nations like Saudi Arabia, or even many of the environmental organizations making up the Climate Action Network. The general sense was that the US demand clashed with the Framework Convention's principle of common but differentiated responsibilities.[10] The NGOs sought a 20 percent cut in industrial country emissions by 2005 and the Europeans a 15 percent cut by 2010, relative to 1990 levels. The United States sought agreement on stabilization of industrial country emissions by 2010 at 1990 levels. With the conference in danger of deadlock and collapse, Vice President Gore arrived on the scene and asked the US delegation to be somewhat more "flexible."[11] The result was a split-down-the middle compromise, in which the United States agreed to cut its emissions by 7 percent by 2008–12 relative to 1990, and the Europeans agreed to cut 8 percent.[12] Within the European Union, widely varying national targets were set, consistent with the general principle of differentiated burdens and tasks.

Gore's role in reaching the agreement was highly controversial, contested, and indeed enigmatic. Why did Gore encourage the US delegation to agree to something he knew could not win Senate approval, at least in the short term? Quite possibly he just wanted to keep the United States engaged in the Framework Convention's negotiating process, however tenuously. If so, he would have been in accord with the European Union's strategic objectives. At no point did the European negotiators countenance an outcome in which all industrial countries except the United States would proceed. And even after adoption of the Kyoto Protocol, when the Senate did in fact decline to ratify, the European Union continued to make every effort to keep the United States involved in the Framework Convention's ongoing negotiations, for fear of an adverse domino effect on other countries in the event of complete US withdrawal.[13]

The irony in all the overblown rhetoric about Gore's role is that the US negotiators got most of what they wanted at Kyoto. In exchange for keeping the United States involved, they obtained agreement on a variety of market mechanisms: the Clean Development Mechanism, a financial device enabling green projects in developing countries to earn and sell credits; Joint Implementation, allowing for industrial country partnerships to reduce emissions, so that dirty projects could be offset by clean ones; and endorsement of the general idea of cap-and-trade emissions trading systems, the most important of which would be Europe's ETS. In what would be a second order of irony, US critics of Kyoto would soon use the practical shortcomings of the CDM, JI, and the ETS—the very instruments they often had favored in theory—to cast further aspersions on the protocol. (Critics ranged from those skeptical about climate change in general to those concerned narrowly about detrimental impacts of Kyoto cuts on the US trade position.)

Kyoto was easy to dislike and denigrate. In short order it became an agreement that a great many Americans loved to hate, from intellectuals to blue-collar workers. One immediate source of unhappiness had to do with how targets were handled for the former Soviet states and the Eastern Europeans. Russia itself and some of its former constituent and client states were mired in depressionlike economic conditions in the mid- and late 1990s, so that their economic output—and greenhouse gas emissions—were well below 1990 levels.

Why did the Kyoto negotiators treat the former communist states so generously? For the United States, the favored treatment of the former Eastern Bloc states translated into the enticing prospect of being able to buy credits in order to meet a Kyoto target it might otherwise have trouble satisfying—the 7 percent cut the Europeans, developing countries and NGOs had elbowed it into accepting. Perhaps, too, US leaders, in their innermost souls, saw the Kyoto mechanism as a kind of back-door foreign aid device that would somewhat compensate for the West's rather niggardly treatment of post-Soviet Russia. The Europeans certainly wanted to win Russia's political support for the protocol—which by the way worked. Because of the way the ratification process was weighted, the Kyoto Protocol took legal force as a treaty with Russia's full accession in 2005.[14]

The special treatment accorded Russia and the East European economies in transition would soon fade from arguments over Kyoto. A problem that would come to loom much larger was the exemption of the developing countries from emission reduction commitments. US negotiators had complained about that exclusion all along, of course. But to the Europeans and their diplomatic allies, the US worries seemed small-hearted and overblown. Surely, in time, if the more privileged countries led the way, eventually the better-off developing countries would get with the program, too. Though total emissions would continue to rise in the short and middle run, the thinking went, eventually the totals would start to come down. Hardly anybody foresaw in 1995–97—including the US negotiators—just how fast China alone would turn into a really big problem.

Scientists and especially economists had never much liked the arbitrariness of the Kyoto emissions targets. "The approach of freezing emissions at a given historical level for a group of countries is not related to any identifiable goal for concentrations, temperature, costs [or] damages," wrote Yale University's William Nordhaus. Furthermore, he said in the same 2005 critique, "The 1990 base year penalizes efficient countries (like Sweden) and rapidly growing countries like Korea and the United States."[15] This is because countries already using energy efficiently would now be asked to become even more efficient, while others continued to get even more inefficient; economically successful economies would have to reign themselves in all the more because of their very success.

Harvard economist Richard N. Cooper declared Kyoto a "flawed concept" that could not be remedied without undermining its fundamental purposes and methods. Even if negotiated emissions limits were derived from business-as-usual growth projections, rather than historical levels, Cooper found that no plausible agreement would result in lower global emissions totals within a reasonable time.[16] He favored instead addressing climate change with a global system of negotiated and harmonized carbon taxes, rather than quantitative emissions limits.[17] Nordhaus lent his support to that notion, arguing that projected costs of carbon limits would always grow more steeply than projected benefits and that curtailing emissions by means of taxes would therefore always be more efficient than by means of numerical targets.[18]

Generally, market-oriented economists tended to believe that the Kyoto Protocol required too much immediate remedial action and that an optimally efficient path would start more slowly with small reductions and then strengthen requirements with time. As a legal assessment summed up this line of thought, "The protocol's model is 'too much, too soon,' imposing sharp near-term reductions that force costly premature retirements of capital stock while leaving uncertain the long-term path of reductions."[19]

Critical arguments of this kind, the discomfort among many NGOs with any kind of market mechanisms, the general perception of an East Bloc giveaway, and the serious unresolved issue of fast-growing emissions from the developing countries combined to feed a general perception—almost universally held in the United States—that Kyoto was doomed to fail. And that attitude would soon produce a stubborn belief that Kyoto had indeed failed although in fact it did not. In a typical assessment, a book published by Cambridge University Press in 2011 concluded, "Very few nations that [had] accepted national emissions limits for [the first Kyoto commitment period] will achieve them."[20] Perceptions of that kind were grounded in unrealistic expectations of what could be expected in the real diplomatic world and a misreading of what Kyoto actually was intended to achieve.

What economists like Nordhaus and Cooper tended to overlook was that the Kyoto targets were intended not so much to achieve a specific goal by a certain time as such but, rather, to set a dynamic in motion that would develop a momentum of its own. Of course the 1990 baseline was essentially arbitrary, and of course the 2008–12 targets owed more to wheeling and dealing—and, in some cases, excessive idealistic ambition—than they did to any rational calculation of costs and benefits. But the basic idea was that as the advanced industrial countries sought to meet their goals, in fits and starts and with much trial and error, the game would catch on. Although this may have seemed a haphazard way of proceeding, the dangers and costs of not proceeding at all would have been much worse.

How well did the industrial countries do in meeting their Kyoto commitments in the initial commitment period that ended in 2012? The question can be answered with precision because, as a legal scholar put it, the protocol has "a more robust compliance system than most other international agreements."[21] Overall, aggregate Annex 1 emissions—not counting the United States, of course, since it had opted out—were down 22.9 percent from 1990 to 2011, according to the IPCC's Fifth Assessment Report.[22] Europe, by far the biggest among the Annex 1 parties, played the biggest part in attaining that result.

When the protocol was negotiated, the fifteen member countries of the European Union were, collectively, after the

United States, the world's second largest emitter of greenhouse gases. As of 2010, the EU-15 was "well on track to achieve [its] commitment . . . of reducing emissions 8 percent compared to baseline levels," according to an assessment by the European Environment Agency.[23] A report from the International Center for Climate Governance confirmed that conclusion two years later. Evaluating overall performance with the protocol in the first three years of the 2008–12 compliance period, the report found that the EU-15 had overachieved its 8 percent reduction target by one or two percentage points. Germany and the UK both exceeded their ambitious targeted cuts of around 20 percent each; Sweden, which was allowed to increase its emissions a couple of percentage points because of its energy efficiency, in fact reduced them by 5 percent. France also achieved a modest cut, despite starting from a relatedly energy-efficient and conservationist position. Of the major EU-15 countries, only two were out of compliance, Italy by approximately 10 percentage points and Spain by 15.[24]

A subsequent report by the European Environment Agency confirmed those findings and then some. Even though Europe's population grew by more than thirty million people between 1990 and 2012, its combined GDP by 45 percent, and its GDP per capita by 36 percent, the E-15 cut greenhouse gas emissions almost twice as much as Kyoto required (by 15.1 percent versus 8 percent). All but three of the fifteen countries exceeded their Kyoto targets, some by wide margins. The most important two factors were indeed the United Kingdom's phase-out of coal and its "dash to gas" and Germany's reorganization of the East's energy sector, but there were many other important elements, too: decreases in emissions in all major economic sectors, except in transportation; decreased emissions of methane and nitrous oxides; especially sharp carbon cuts in manufacturing and construction; improved energy efficiency and conservation in both residential and commercial buildings; ever-greater reliance on renewables; and switching to lower-carbon fuels.[25]

Cynics may say that Europe succeeded only because certain key countries got commitments that they knew they could easily meet. But such an argument is slippery and dangerous, as the emissions trading pioneer Sandor commented, in connection with the spectacular success of the US acid rain program (see chapter 3). The plain fact of the matter is that the European Union succeeded because some of its leading members were absolutely determined to see it succeed and adopted policies to achieve success. This goes even for Britain's coal phase-out and Germany's rationalization of the East's power sector: The United Kingdom's shift to natural gas would not have occurred without adoption of rules favoring competition in electricity, and Germany would not have been successful had it not opened its eastern states to foreign investors; Sweden's national utility, Vattenfall, was allowed to buy much of the East's electric power infrastructure and proceeded to improve it dramatically.[26]

Globally, according to the Climate Governance Report, twenty-one parties to the Kyoto Protocol were in compliance with it as of 2010, and seventeen were out of compliance.[27] Super-overachievers were mainly in the former East Bloc and included Poland, Russia, and Ukraine. They roughly balanced and compensated for the egregious underperformance by Canada and poor performance on the part of Austria, Spain, and Japan. Japan, the only underachiever among the world's very large economies, had greenhouse gas emissions that were about 5 percentage points above target—not a good performance, but not dreadful either.

All of this leaves out of account, of course, the major Kyoto delinquent, the United States, which repudiated the protocol and therefore does not count in an assessment of how parties did in meeting commitments. But for the record, US greenhouse gas emissions in 2008–11 were 20 percentage points above the 1990 Kyoto baseline. Thus, Obama's 2009 pledge of a 17.5 percent cut by 2020 as compared to 2005 would not get the United States even back to the baseline year, let alone the 7 percent reduction from baseline that the protocol would have required. To many or most American citizens Kyoto may have seemed an absurdity, but as far as the Europeans and their diplomatic allies in the third world were concerned, "the U.S. attempt to unilaterally scrap an agreement negotiated over almost a decade and agreed [to] by 180 states was judged a 'provocative' and 'brutal' diplomatic act," one scholar wrote.[28]

Is the Kyoto Protocol, at bottom, despite its successes, an absurdity? And must its underlying and animating principle, the idea of common but differentiated responsibilities, be discarded in favor of something else? It may be instructive in this context to consider another international accord that everybody has loved to hate, the Nuclear Non-Proliferation Treaty (NPT) of 1970.[29] In that treaty, countries without nuclear weapons agreed to renounce their acquisition in exchange for a vague commitment on the part of the nuclear weapons states to engage in a long-term effort to ultimately rid themselves of atomic arms as well (a promise, by the way, that none of the nuclear weapons states had, at the outset, any honest intention of keeping). Conceptually, the NPT and Rio Framework Convention are almost exact mirror images of each other: In the one, the world's developing countries plus some important advanced industrial countries agree to forego obtaining nuclear weapons, for the sake of making the world marginally safer in the short run while establishing the conditions of making it much safer in the long run. In the Rio and Kyoto follow-on treaties, the advanced industrial countries take action to cut emissions right now, with the aim of somewhat reducing emissions and associated risks in the short run, to initiate a process that will lead to much bigger payoffs in the longer run.

The Non-Proliferation Treaty, like the Kyoto Protocol, always has been subject to lots of well-founded objections. From the time it first was under discussion and negotiation, non- and near-nuclear countries, with India in the diplomatic lead, argued that the treaty would "disarm the unarmed while leaving the armed free to keep arming." To overcome that admittedly formidable objection and entice nonnuclear countries into the treaty, there was a deal. Under the basic terms of the treaty, non-nuclear weapons states were to have unfettered access to all "peaceful nuclear technology" as long as they agreed to regular international inspections, designed to give timely warning of any diversion of nuclear material from peaceful to military applications. As critics have complained ever since, the treaty enabled weapons-prone countries to acquire all the technology and materials needed to produce atomic weapons and then break out of the inspections regime to quickly go nuclear. There were bound to be such cheaters, and yet the NPT contained no provisions spelling out how they were to be dealt with even if they were caught red-handed.

Such was the atmosphere of intellectual pessimism and political cynicism that attended the birth of the NPT that the tendency ever since has been to declare the expected failure a failure indeed. Yet the plain facts of the matter are radically at variance with that perception.

At the time the treaty took force in the early 1970s, there were five recognized nuclear weapons states plus one that was undeclared but well on its way (Israel). In a 1976 article that appeared in a leading US journal of international affairs, "Spreading the Bomb without Quite Breaking the Rules," political scientist Albert Wohlstetter—an eminent figure in his day—predicted that within fifteen years some seventy-five countries would have enough separated plutonium obtained from normal reactor operations to make nuclear weapons.[30] By the end of the century, he clearly implied, every Tom, Dick, and Harry would be in a position to develop nuclear weapons, and many would in fact do so.

Today, forty-five years after conclusion of the Non-Proliferation Treaty and nearly four decades after Wohlstetter issued his dark forecast, only four additional countries have acquired nuclear weapons, and only two of them have come to be recognized de facto as nuclear weapons states, India and Pakistan. Israel's nuclear arsenal remains undeclared and is not accepted as legitimate in much of the world; North Korea's is considered flatly unacceptable and, at least in principle, reversible. The NPT itself has gained nearly universal adherence, so that the nonnuclear norm has become in effect an element of international common law. Without general acceptance of that norm, it is scarcely plausible that a number of countries attempting to go nuclear would have been stopped or very nearly stopped in their tracks. Soon after the treaty's conclusion, the United States got client states like South Korea and Taiwan to terminate nuclear-weapons-oriented activities; later the United States would put the screws on Argentina and Brazil. The dying apartheid regime of South Africa was persuaded at the end of the 1980s to dismantle fully built atomic bombs that it was ready to test and instead go nonnuclear.[31] Immediately following the first Gulf War, it was discovered that Saddam Hussein was developing an elaborate industrial infrastructure to develop nuclear weapons, which was completely dismantled under the supervision of United Nations inspectors.[32] Muammar Gaddafi of Libya was muscled into reaffirming his NPT commitment in 2003 after he was caught smuggling critical technology into the country, having acquired a bomb blueprint from the Pakistani nuclear racketeer A. Q. Khan.[33] Currently, North Korea, having crossed the weapons threshold, is under acute pressure, including from its patron China, to reverse course. Iran, which has been intent since the beginning of the twenty-first century on acquiring all the wherewithal for a bomb, has had to reverse course as well: Because of the nonproliferation principle, it was subjected to devastating economic sanctions, which not only the European "frontline states" agreed to—the states that initially confronted Iran over its twenty-year violations of its NPT commitments—but also China and Russia.[34]

In light of that well-documented history, we know what would have happened in the absence of the Non-Proliferation Treaty about as well as it is possible to be sure of any counterfactual proposition. Without the treaty, first Germany and Japan would have likely gone nuclear—keeping them nonnuclear was after all the immediate aim of the superpowers and what got them talking seriously about a nonproliferation treaty in the first place. Asian tigers like Taiwan and South Korea would surely have followed suit. South Africa and Iraq would be nuclear weapons states, and most likely, Argentina and Brazil would have gone nuclear, too.[35] In a world like this, not only would the daily risk of nuclear war be significantly higher, but any prospect of eliminating the nuclear scourge would be absolutely unthinkable.

The Nuclear Non-Proliferation Treaty has in fact been so successful and is so well established in world diplomatic opinion as successful that it was made permanent in 1995, rather than having to be renewed every five years. Though there are still some hold-out and delinquent countries, membership in the treaty has come to be almost universal.

To return to Rio and Kyoto, what will the consequences be if the principle of common but differentiated responsibilities is abandoned and efforts to renegotiate a concerted global program of greenhouse emissions fail? The likely results are by now so familiar to most educated people that it is tiresome to repeat them: rising sea levels and increased flooding of coastal cities and deltaic plains; more frequent, prolonged, and severe droughts in the world's drier agricultural areas, and more frequent and intense rainfall in the wetter regions; more extreme and severe weather generally; and, ultimately, the possibility of an abrupt change in overall climate dynamics, or a concatenation of regional climate disasters, leading to a world-scale catastrophe.[36]

So familiar is the usual litany of damages that major world authorities have been grasping for new ways to restate just how dire the outlook really is. The IPCC, in its latest assessment, equated the goal of stopping warming at 2°C to the total accumulated combustion and emission of one trillion tons of carbon; more than half that amount already has been pumped into the atmosphere since the Industrial Revolution began, around 1750, leaving us with little more than two-fifths the total to spew in the decades to come. If business continues as usual, without coordinated international remedial measures, we shall burn our trillionth ton of carbon on November 25, 2040, according to an Oxford University calculation.[37]

The UN Environment Programme, in its 2013 report, preferred to state its warning in terms of the high price of delayed action on climate. The agency reported, as one journalist wrote, "that even if nations meet their current emission reduction pledges [per Copenhagen], carbon emissions will be 8–12 gigatonnes above the level required to avoid a costly nosedive in greenhouse gas output."[38] In other words, if the world is to have any chance of meeting its climate objectives, unless it takes more concerted action quite soon, it will have to take *much* more concerted action not much later. Otherwise we will be heading for a future in which the world could be 4°C or even 6°C hotter.

Perhaps the most vivid recent warning came from a group of scientists who calculated the year in which, for various world cities, the average year would be hotter than *any* year recorded between 1860 and 2005 (the period for which detailed temperature records are considered highly reliable). For Manokwari, Indonesia, that year could be as soon as 2020, and for Kingston, Jamaica, it would be 2023. The "new normal" would arrive in New York City and Washington, DC, in 2047 and in Orlando and Phoenix in 2046. Moscow will start seeing the new normal in 2063, and Anchorage, Alaska, only in 2071. "In all, the scientists found that between 1 and 5 billion people [eventually] would be living in regions outside such limits of historical variability."[39]

9

Copenhagen

In December 2009, representatives of all the world's nations, including some sixty heads of state and perhaps forty thousand individuals from nongovernmental organizations, assembled in Copenhagen for what was both meant and widely expected to be a breakthrough climate conference. Negotiators faced two challenging tasks. One was to reach agreement on a next phase of greenhouse gas reductions pursuant to the 1992 Framework Convention and Kyoto Protocol. At Kyoto, the advanced industrial countries had promised to make specific cuts in greenhouse gases by the end of the following decade; at issue now was what they should do in the second decade of this century. The second and even more difficult task was to reformulate a general philosophy about how climate negotiations would go forward, in particular how to bring large, rapidly industrializing countries like China, India, and Brazil into a general program of greenhouse gas emissions limits. Despite the challenges, the prevailing mood in Copenhagen was optimistic, first and foremost because an exciting new president had taken office in the United States that year and had immediately signaled, even before his formal inauguration, that climate change would from now on be a central concern in the White House.[1]

With a remarkable cast of characters assembling in Copenhagen, speculation centered on whether this fifteenth Conference of Parties—COP 15—would be a "good COP" or a "bad COP." Almost everybody in attendance fervently wished for it to be a good one, and just about everybody assumed that the United States would find a way to start playing ball again. President Obama would arrive fresh from being awarded a Nobel Peace Prize, but he was by no means the only flashy or important person on his way to Copenhagen. Also registering high on the global Richter scale of political celebrity was Brazil's president Luiz Inácìo Lula da Silva, who had consolidated his country's nascent democracy on the basis of a rebalanced social contract between poor and rich. Representatives of significant negotiating blocs included the Sudanese diplomat Lumumba Stanislaus-Kaw Di-Aping, chairman of the Group of 77 non-aligned developing countries; Maldives president Mohamed Nasheed, principal spokesman of AOSIS; and European Commission president José Manuel Barroso, representing the group of countries that had been taking climate change most seriously and that had been implementing the Kyoto Protocol most successfully. Finally there were those in the spotlight by virtue of what and who they stood for: China's premier Wen Jiabao, France's Nicolas Sarkozy, Germany's chancellor Angela Merkel, Indian prime minister Manmohan Singh, South African president Jacob Zuma, and the United Kingdom's Gordon Brown. None of these was necessarily popular with everybody at home or abroad, but most were substantial political personalities.

The supporting and kibitzing cast of global activists also commanded attention. South Africa's archbishop Desmond

Tutu was there, bringing his formidable moral authority, acuity and eloquence to bear on what many were calling the defining challenge of the twenty-first century. A big display in Copenhagen's downtown pedestrian mall let passersby know that Brad Pitt cared about the world's future. At a downtown convention center where nongovernmental organizations held a parallel conference, speakers included the ubiquitous climate activists Bill McKibben and George Monbiot, from the United States and Great Britain respectively. McKibben had just launched 350.org, a grassroots movement dedicated to getting the atmospheric concentration of carbon dioxide back down to 350 parts per million from close to 390 ppm, where it was hovering at the time.

It was a measure of how far educated public opinion had come in the previous decade that people actually knew what McKibben was talking about. Inside the Bella Center on the outskirts of the city, where COP 15 had convened, national delegations were arguing about whether the world could afford to let the concentration of CO_2 in the atmosphere reach 550 ppm in the 21st century—representing an estimated warming of three degrees Celsius—or whether it would be necessary to pull out all the stops to keep the level at 450 ppm, for an increase of two degrees Celsius. In the preceding years, a fair amount of scientific consensus had developed that beyond 450 ppm global warming might become truly dangerous (see chapter 1). But McKibben and his followers took the position that the world was already in the danger zone: Inasmuch as the level of carbon dioxide already was almost 50 percent above the preindustrial level, they argued, it was urgently necessary to start ratcheting down concentrations right now, whatever the cost. In the Bella Center, delegations from endangered island states and sub-Saharan African nations adopted the same line. Articulating their position, Archbishop Tutu sent a letter to heads of state and Christian leaders saying that since Africa was expected to warm by more than the global average, "A global goal of about 2 degrees Celsius is to condemn Africa to incineration and no modern development."[2]

A stirring moment occurred on the Sunday afternoon separating the two weeks of the conference, December 13, when activists joined McKibben and Archbishop Tutu at a downtown church to hear its bells toll 350 times. But that event turned out to be something of a spiritual and emotional high point, because it was already clear that COP 15 was not going according to script. Violent clashes were occurring in the streets between demonstrators and police, leading to arrests. The conference itself was in full revolt, with representatives of the third world denouncing a leaked "Danish text" of a final agreement.[3] Diplomats from the world's less advantaged nations thought the draft threatened to depart from the Rio principle of common but differentiated responsibilities and the Kyoto program, which assumed that developing countries would be required to make emissions cuts only after the industrial countries had first executed a twenty-year, two-phase program of sharp reductions.

Meanwhile, outside the conference venue, thousands of accredited people stood in long lines, braving Denmark's wintry chill and gloom, unable to gain admission.[4] Many had come from around the world to take part in the conference as observers and lobbyists for nongovernmental organizations. It turned out that the conference organizers had authorized forty thousand people from NGOs to attend, but now Danish security said the Bella Center could accommodate only fifteen thousand. Because of the over-crowding, and because of splits in the NGO community between those favoring and those opposing civil disobedience, the NGO representatives ended up largely shut out and disenfranchised for the remainder of the conference.[5]

Yvo de Boer, the designated chief operating officer of the Rio-Kyoto process, had said the Copenhagen conference would be deemed a success only if it produced significant and immediate action, starting the day the conference ended.[6] But now, as the second week began, any such outcome was gravely in doubt. Conference chairperson Connie Hedegaard had spoken with cautious optimism in the months before about getting countries like China to state when their greenhouse gas emissions would peak.[7] Now, having conspicuously failed to obtain such a pledge and in the face of the developing countries' rebellion, she had to relinquish the chair to Denmark's prime minister. A general mood of uneasiness and restlessness prevailed as attendees awaited the arrival of the really big players.

Despite the unpromising circumstances, on the afternoon of Wednesday, December 16, Senator John Kerry, representing the Foreign Relations Committee, delivered an upbeat talk in the conference center, where he was warmly received by delegation members and NGO representatives. After years of "delay, divide, deny" on the part of the United States, Kerry said that Americans now recognized that there was no being "half pregnant" on climate change; you either understood the science or you did not. The new administration got the message and meant business, he continued. Already it had committed eighty billion dollars to clean energy, more than half the US states were party to carbon emissions trading systems, and some one thousand mayors had taken initiatives to help meet Kyoto targets. A House bill set a goal for US greenhouse gas emissions, and the outlook was good for Senate passage of a cap-and-trade bill, he thought; what was more, if the Senate did not act, well then, the Environmental Protection Agency would regulate carbon under authority of the Clean Air Act, as the Supreme Court had allowed.[8] But if the United States was to take really effective action on climate, said Kerry, its political leaders had to be able to tell their constituents that they would not lose their jobs because of China's or India's refusing to make similar commitments. That is, the United States was ready now to play ball, but the big, fast-industrializing countries like China had to get in the game as well.

Implicitly, members of the audience seemed inclined to take Kerry and the Americans at their word: They thought that although the United States would and could not join in the Kyoto process as such, it was ready to take a supportive role in the talks. But that kind of assumption about US intentions and resolve may have been overly optimistic. Although it would take a year for Kerry's expectations about US domestic politics to be shown definitively wrong, with the administration's sharp setback in the November 2010 midterm election, it took only a week or ten days to learn that the conditions the United States was setting for China and India—transparent emissions accounting, independent verification, and emission pledges— were not going to be satisfied at Copenhagen. The Chinese delegation had a phobic reaction to any suggestion that the country's greenhouse gas performance should be independently monitored and verified. As for Hedegaard's notion that China might declare a year in which its emissions would peak, China reacted so hostilely to the very idea of emissions pledges that it tried to get all mention of them removed from the proceedings. Chancellor Merkel did not conceal her incredulity at China's notion that those actually making pledges should not be allowed to state them in public.

By the night of Thursday, December 17, the conference was at an impasse. At that point, President Obama and Secretary of State Hillary Clinton barged into a late-night meeting that the leaders of China, India, and South Africa were having among themselves. In the next hours, they hammered out the outlines of what came to be known as the Copenhagen Accord, a semiformal agreement "acknowledged" but not formally adopted by the whole conference.[9] Basically it said that parties, including both industrial and fast-growing, less-developed countries, would submit pledges concerning future greenhouse gas emissions. Future hard commitments would be negotiated on the basis of "common but differentiated responsibilities," the core principle of the Rio treaty that the developing countries saw as imperiled when the "Danish draft" had circulated the week before. By 2020, the rich countries would provide one hundred billion dollars annually to the less advantaged countries to help them minimize and cope with climate change. Regarding the key issue of second-phase Kyoto commitments, for the period 2012–20, the accord was silent, neither endorsing nor repudiating the Kyoto process.

Seen from the perspective of the official US negotiating position, the Copenhagen Accord was a pretty good deal. In the months before the conference, the chief US negotiator Todd Stern had indicated clearly that the United States would not agree to "do Kyoto" under any circumstances, and he firmly rejected any notion that the rich countries owed poor countries "reparations," because of past greenhouse gas emissions.[10] In the accord, the United States gave a little ground on emissions commitments, only vowing with others to make an actual pledge. It got the developing countries to agree to biennial reporting of emissions. And it was not criticized for not reengaging in the Kyoto process, which the accord said nothing about.

Where the United States appeared to concede the most, ironically, was in the area of aid for developing countries: Though the accord did not suggest or imply that rich countries were morally indebted to the poor, the industrial countries were committed to put up a prodigious amount of money not far down the road. Whether it represented a reparation for past greenhouse gas emissions, one hundred billion dollars per year, if it ever materialized, was nothing to sneeze at.[11]

Seen from a global perspective, however, the Copenhagen Accord fell far short of what had been called for. The Bali Action Plan, adopted at COP 13 two years previously, said that a Copenhagen "decision" should include "a shared vision for long-term cooperative action," "quantified emission limitation and reduction objectives, by all developed country Parties," "nationally appropriate mitigation actions by developing country Parties," "cooperative sectoral approaches and sector-specific actions," and ways of engaging "private sectors and civil society," among other things. The Copenhagen Accord had little or none of that, though it did recognize the 2°C warming limit as a reasonable objective—an important outcome.[12]

Though China and India declared themselves more than happy with the accord, which did not require much of them, not everybody involved in forging it was content with it. In a short but impassioned address to the plenary conference, on the next and last day, Brazil's president Lula expressed his "frustration" with the meeting. Instead of real action on climate change, what he had seen was bargaining in targets and a false notion that in providing money for climate relief, the privileged industrial nations were delivering alms. The former union organizer said that although it had been thrilling to sit in a room with global political celebrities in the middle of the night, in the end the conference had been like a labor-management negotiation gone awry: The parties to the talks, like short-sighted trade unions and enterprises, lost sight of their common responsibilities and acted as "if it is just about money."[13]

The journalist Mark Shapiro, who was covering COP 15, said that Lula had lost patience watching a closed-circuit transmission of Obama's remarks to the conference.

> The hope in Copenhagen was that the United States and other developed countries would commit to specific emission limits, and major developing countries like China, Brazil, and India would commit to less specific but internationally verifiable reductions. You could almost feel the air being sucked out of the catacomb of corridors in the negotiating center as Obama retreated from any emission reduction commitment by the United States, and threw some of the responsibility for his position back on China's unwillingness to accept international emission monitors operating within its borders. Lula, according to several members of the delegation who were present during the speech, was disappointed and infuriated at Obama's dodge and tossed out the speech in which he'd planned to celebrate the hoped-for US commitment to a global accord.[14]

Lula could have reacted differently. Rather than grumble about money hunger, Lula could have styled himself and his country as leaders in greenhouse gas avoidance, so as to put pressure on those less successful. Brazil, with an electricity sector that draws heavily on carbon-free hydropower and transportation that runs to a great extent on biofuels, has some of the lowest per capita greenhouse gas emissions in the world

(2.2 metric tons).[15] Another successful country is France: In electricity, it went all nuclear in the 1960s and 1970s and always has discouraged automotive transportation. At Copenhagen, Brazil and France could have waved nagging fingers at the United States, which among the world's big economies has the highest per capita greenhouse gas emissions (17.6 metric tons), and other delinquent advanced industrial nations.[16] But French president Nicolas Sarkozy, usually so brash and abrasive, kept a low profile. And Lula preferred to cast himself as spokesman for the world's poorest and most climate-endangered states, not as a carbon-cutting pioneer.

By the end of the conference's second week, those most immediately endangered by climate change—from small island states of the Pacific to big, low-lying states like Bangladesh—knew that they were not going to get commitments from the advanced industrial countries to make further sharp reductions in emissions. At least some of their representatives, though, were coming around to the view that the proposed process of voluntary pledge making was not completely worthless. In effect, that process drew on the various pledges that a number of major countries and regions had made coming into the meeting, pursuant to a suggestion the US delegation had made at Bali.[17]

In the following years, as the pledges were formalized in keeping with the Copenhagen Accord, they would come to be a significant part of the domestic political chemistry in pledging countries. Advocates of aggressive climate action could always refer back to them as a matter of national honor and thus, in a way, beyond debate. President Obama would make notable mention of his Copenhagen pledge in mid-2013, when he unfurled his administration's Climate Action Plan.[18] Still, nobody pretended that the pledges had the same weight as treaty commitments to make reductions would have had. The

Copenhagen Accord was by general consent better than nothing, but how good—or how disappointing—was it, in terms of what the global climate crisis demanded and in terms of what global politics could have permitted?

Clearly the Copenhagen Accord did not deliver the most important "deliverables" that de Boer had said would be the test of whether COP 15 was a success or failure. In the years leading up to the conference, the Rio Convention's executive secretary listed the critically needed products of the conference: ambitious midterm emissions reductions by major industrial countries; clarification by the main developing countries of what they could do to limit emissions on a similar schedule; financial commitments to help less-well-off countries with the adverse effects of climate change; and agreement on future governance structures to regulate ongoing efforts.[19]

De Boer, a sober-minded realist and not a starry-eyed idealist, obviously had expected more of the agreement and had thought that a better agreement was politically achievable. In particular, he had seen reason for optimism in a new spirit of cooperation he thought he sensed between the US Senate and the American delegation. "I think that a major shortcoming of Kyoto was that the official delegation came back with a treaty they knew was never going to make it through the Senate," he told the *Guardian.* "This time I have the feeling that communication is much stronger, that the Senate Foreign Relations Committee, through John Kerry, is really expressing what they feel needs to be done in Copenhagen."[20] If de Boer expected more, it almost goes without saying that the forty thousand individuals representing NGOs—including all the world's leading environmental organizations—expected much more.

Generally, the Copenhagen Accord "was vague or weak on key points," as two legal scholars later would observe. "Moreover, objections from just five nations blocked the accord from formal adoption, so its status even as a basis for future negotiations is uncertain. In sum, despite modest favorable signs before and at Copenhagen and current high attention to climate change, it remains unclear whether leaders of major nations are willing to act strongly enough to address the problem, or whether the current international negotiation process is able to motivate and coordinate such action."[21]

Could much more have been achieved? At least in theory it could have. Suppose, for example, that President Obama had arrived a little earlier and had made a point of meeting first with the leaders of the most important European countries. He might have started by acknowledging their leadership in cutting greenhouse gas emissions and by expressing regret over the unconstructive and almost obstructive role the United States had played for the previous twelve years. He might then have promised, on his own authority, to make radical cuts in US greenhouse gas emissions in the next seven years—and to make them as sharp as necessary to get China and India into the game. If necessary, Obama could have reminded the Europeans of the singularly cumbersome US treaty ratification process. As a political vote counter among political vote counters, he might even have explained that in order to win Senate support for any new climate treaty, he would not only have to hold all fifty-nine Democrats and Independents but win over eight Republicans— an absolute impossibility. So there would be no question of the United States formally rejoining the Kyoto process or indeed agreeing to any kind of ambitious climate treaty.

Having won over the European leaders with his sincerity, charm, and eloquence—to continue with this counterfactual thought experiment—Obama and those leaders would now go to see the leaders of China, India, and Brazil. The prime minister of Britain and chancellor of Germany, as the leaders of the two countries most successful in sharply cutting their greenhouse gas emissions, would reiterate what Obama had just told them. Obama would then confirm their account, specifying that he would personally guarantee by executive agreement that the United States would make sharp emissions cuts in parallel with the second-phase Kyoto cuts and that he would lend his full personal support to the Kyoto 2 process. Thereupon, under the combined moral pressure of the United States and Europe—and the implied material threat of trade reprisals—China and India would agree to measures that would be more ambitious and specific than those embedded in their actual Copenhagen pledges.

Though all that might be dismissed as an idle fantasy, arguably something like that is pretty much what most participants at Copenhagen, consciously or unconsciously, expected to see happen. In hindsight, now that we know Obama much better, it is obvious that the highly methodical and risk-averse president would not likely have chosen the alternative path. And perhaps no other path was really feasible. Alf Wills, lead climate negotiator for South Africa, believes that so much time had elapsed since the United States had turned its back on the Kyoto Protocol, and so much else had happened in the meantime, that "a rebalancing of commitments had to take place."[22] In any case, in taking the course Obama picked, he not only delayed conclusion of a new comprehensive climate agreement by at least six more years, he also rendered the second Kyoto commitment period virtually meaningless, so that there would be no firmly agreed-upon internationally coordinated climate action for more than a decade. At a time when the visible effects of global warming were already getting much more alarming than most experts had been predicting, this loss of time in organizing concerted action was a big loss indeed.

In assessing the shortcomings of the Copenhagen conference, two schools of thought immediately emerged. One found expression in a blog written from Copenhagen by Mark Lynas for *The Guardian*. Lynas saw the outcome as a flat disaster and put the blame squarely on China, for making Obama's job impossible. "China wrecked the talks, intentionally humiliated Barack Obama and insisted on an awful 'deal' so western leaders would walk away carrying the blame. How do I know this? Because I was in the room and saw it happen."[23] A month later, writing in the *New Statesman*, Lynas expressed wonderment that many climate campaigners rejected his claim that "it was the developing world—primarily China and India—that [had] destroyed the putative 'deal.'"[24] Lynas peremptorily rejected developing country arguments that carbon emission was a prerequisite of economic growth, that historical emissions by the rich countries somehow conferred "an equal right to pollute" on the part of the poor countries, or that there were any valid reasons for referring to so-called climate justice as an excuse for carbon pollution. Lynas asserted that to the extent the rich countries owed the rest of the world anything at all, they were amply meeting that obligation with the hundred-billion-dollar climate fund they agreed to create. The *Guardian*'s on-the-spot Copenhagen blog—presumably Lynas's work, too—credited US Secretary of State Clinton with "saving the talks from collapse" by acquiescing to the fund.[25]

A diametrically opposed school of thought held that the United States had brought the talks to the brink of collapse. David Corn and Kate Sheppard, writing on *Mother Jones* magazine's website, described the agreement Obama had forged with China, India, South Africa, and Brazil as a sly and slightly diabolical maneuver. In essence, as they saw it, the United States had "sidestepped the official proceedings and found a way to separate major developing countries from poor ones—while skating past European desires for a more comprehensive and binding agreement."[26] The result, in the words they quoted of Erich Pica, president of Friends of the Earth USA, was an accord that was not strong, just, "or even real."

The one major party to the Copenhagen talks that got little or no blame at the time was the European Union or the major European powers. In the cold light of hindsight, that is rather curious. In the five years following the Rio Earth Summit of 1992, it was the European Union that struggled hardest to negotiate an agreement committing the advanced industrial countries to emissions cuts. On adopting the Kyoto Protocol in 1997, Europe worked doggedly to keep the United States engaged in climate negotiations, despite US repudiation of the protocol. And among the industrial economies, Europe had been most successful in meeting Kyoto commitments, having cut emissions almost twice as much as the protocol called for.

Despite that formidable moral position, European conduct at Copenhagen was weak in every possible way, as the legal scholar Cinnamon Carlarne has detailed. First off, "the Danish faltered in their conference leadership abilities." Sarkozy failed to advance the common European agenda, drifting from "the party line." With the German chancellor failing to make a powerful showing and England's prime minister dominating headlines with rhetoric (but not real action), "the EU bloc as a whole was sidelined in negotiations."[27]

So why did the Europeans not seek to capitalize on their moral advantage at Copenhagen and press the United States and China much harder? Three major factors surely were in play. First, there was the inclination to trust and not insult the new American president, who after all was such an immense improvement over his predecessor in terms of climate rhetoric and actual energy policy. Second, there may have been an element of negotiation fatigue: For almost two decades, the Europeans had been arguing almost nonstop with the United States over the allied questions of mandatory emissions cuts and developing country obligations. Maybe they just wanted to take a break. Third, and surely most important, the world economy was teetering on the edge of depression at the end of 2009. The crisis was proving to be particularly long and obdurate in the Eurozone. So, if the United States and China were dead-set on sidetracking all talk of costly new mandatory emissions cuts, perhaps the Europeans were quietly content to let them do so and take the heat.

Taking all the preceding lines of critique into account, to the extent that Copenhagen failed to meet expectations, it can be said to have been the result of two sins of commission (America's and China's) and one sin of omission (Europe's). But in retrospect it also is obvious that expectations were too high to begin with. Naomi Klein, another journalist at Copenhagen, narrated a telling anecdote about a young climate justice activist who had been "the picture of confidence and composure, briefing dozens of journalists a day on what had gone on during each round of negotiations." But when it was all over and, as Klein put it, "the pitiful deal was done," he fell apart during a meal and began to sob. "I really thought Obama understood," he kept repeating, Klein wrote.[28]

Not everybody was so naive. Alden Meyer, point man on climate for the Union of Concerned Scientists, says that he and other leading NGO lobbyists were clear well before Copenhagen that the conference would not live up to its billing. Shyam Saran,

India's climate envoy at the time, noted that in October 2009, at a preparatory conference in Bangkok, the United States was making clear that it would not accept binding emissions cuts and Europe was drifting toward support for an alternative to Kyoto that would be attractive to the United States. "With the US having set the bar very low, it was inevitable that other developed countries would all gravitate to the lower common denominator," Saran wrote later.[29]

Surya Sethi, a Core Climate Negotiator for India's government, also saw where things were—and spelled it out publicly weeks before the conference began. In a videotaped interview, he recognized that the European Union was caving into the US position and the result would be a stalemate at Copenhagen on the key question of industrial country commitments.[30]

Sethi had a further complaint, but that was about India itself. Before and during the Copenhagen conference, he urged India's government to assume political leadership of the world's poorest people and countries—to reassume leadership, if you will, of the nonaligned movement that Nehru had cofounded.[31] But India's top politicians and diplomats preferred to align themselves with the more prestigious BRIC/BASIC bloc, which led to the anomalous result that China—the world's largest economy, with per capital carbon emissions approaching Europe's!—was able to present itself as the G-77 leader. Just as shameful, arguably, was the attitude of the developing countries generally. Though they were calling for the strongest possible action and their advocates were demonstrating in the streets, yet they still took the position that only the world's most advanced industrial countries—basically North America, Europe, and Northeast Asia—needed to take stronger action right away. They did nothing to contribute to the pressure that industrial countries were trying, however feebly, to put on China.

Copenhagen, in a nutshell, epitomized all that had been wrong in two decades of climate diplomacy. It was not the complexity of the issues or the unmanageability of the process. It was diplomatic pusillanimity on the part of every major participant, from the European Union and the United States to China, India, and the G-77.

10

The Road to Paris

By all accounts the preparatory work for Copenhagen had been inadequate, major players came into the conference without a clear understanding of each other's positions, nongovernmental organizations were almost literally shut out, the Europeans allowed themselves to be coopted by the Americans, the G-77 and AOSIS were coopted by China, and India fiddled while everybody else fumed. In the end, none of the participants acquitted themselves especially well. It was, said Richard Kinley, the longtime Number Two at the Framework Convention's secretariat in Bonn, "in many ways our worst nightmare." But it also turned out to be, he immediately added, "the most successful failure one could have imagined."[1]

That was because, although nobody came out of Copenhagen feeling good, just about everybody decided to make the best out of what seemed to most a not-good situation. At the next Conference of Parties, in Cancún, many countries began to formalize the pledges that some had made informally coming into Copenhagen. The two-degree goal was formalized and agreed upon. The year after that, at COP 17 in Durban, an action plan was adopted that called for an ongoing negotiation to culminate in a new major climate accord with legal force, which was to be concluded at Paris in December 2015 and would govern parties' actions starting in 2020.[2] In Warsaw, COP 19 gave further definition to the climate action pledges, with the term Intended Nationally Determined Contribution, and called on the task force established in Durban—the Ad Hoc Working Group on the Durban Platform, or ADP—to specify the kind of information that should be included in INDCs, so that they could be compared meaningfully.

In the final two years leading up to Paris, to judge from what seemed to be going on in the formal negotiations, progress was, yet again, halting. It was already clear at the Warsaw Conference of Parties that China and India would not agree to a process involving intensive critical review of INDCs.[3] The final conference before Paris, in Lima, produced an enormously long draft agreement that was filled with technical quibbling and did not set a tight-enough deadline for submission of INDCs to allow for meaningful review (or so it seemed at the time). Starting six months before Paris, senior negotiators indicated to journalists that more headway was being made behind the scenes than was apparent in the open talks, but that remained to be seen.[4] Meanwhile, however, experts were developing a body of sophisticated thought about what a successful treaty might look like. Particularly noteworthy were two papers, Expectations for a New Climate Agreement, by a pair of MIT researchers, and "Building Flexibility and Ambition into a 2015 Climate Agreement," by Daniel Bodansky and Elliot Diringer.

The two MIT scholars, Henry D. Jacoby and Y.-H. Henry Chen, were writing as technical experts on climate change and industrial policy; each has degrees in both engineering and economics. Jacoby, the much more senior of the two, is a retired professor and former codirector of the MIT Joint Program on the Science and Policy of Global Change who had served in a variety of impressive positions at both Harvard and MIT, including chairman of the MIT faculty. Bodansky, who has been mentioned in previous chapters, was briefly the senior US climate negotiator, straddling the administrations of Bill Clinton and George W. Bush. (By his estimation, he got the job almost entirely on the strength of a legal analysis he had published of the Rio Convention.)[5] Since leaving government, he has been a law professor at Arizona State University and has written frequently on the technicalities of climate agreements, often publishing on the important site of the Center for Energy and Climate Solutions, C2ES. His coauthor, Diringer, is a researcher with C2ES.

Jacoby and Chen proceeded from the assumption that the Paris agreement would be based on the two-step process of pledge and review as specified at Durban and Warsaw, and not on a Kyoto-like system of mandatory quantitative emissions cuts. Focusing on a target date of 2030, they compared the totality of pledges that might reasonably be expected to come out of Paris with a reference emissions scenario in which countries do no more than they promised to do at Copenhagen.[6] They found that the expected Paris outcome would yield total carbon-equivalent emissions in 2030 somewhat over 60 billion metric tons—about 10 billion metric tons lower than the Copenhagen reference case but at least 10 billion metric tons higher than a window consistent with 2°C warming (roughly 40 to 53 billion metric tons).[7] In 2050, the Paris outcome would be about 20 billion metric tons lower than the Copenhagen reference case but 20 billion metric tons higher than the 2°C window. In short, their analysis concluded "that these international efforts will indeed bend the curve of global emissions. However, our results also show that these efforts will not put the globe on a path consistent with commonly stated long-term climate goals."[8] Obviously, as Jacoby and Chen concluded, their analysis raised questions about the adequacy of the INDC pledge-and-review process as it was then being executed and suggested strongly that the process would need to be strengthened in the future.

Their conclusions about China and the developing countries also were noteworthy. Their study, published a month before the "breakthrough" US-China climate agreement described in chapter 5, found that Chinese emissions would stabilize around 2025–30 "under the policies and measures assumed."[9] That suggested that the Chinese 2030 peaking pledge represented little more than an acknowledgment of what was already planned or foreseen. As regards the developing countries, their Paris scenario assumed that in the coming decades, the developing countries would account for accelerating shares, relative to the industrial states, of mandated renewables, automotive fuel efficiency improvements, and household energy efficiency.[10]

Bodansky and Diringer, in their June 2014 analysis of a prospective Paris agreement, likewise surmised that the aggregate Paris pledges would not "put the world on a pathway to meeting the 2°C temperature limit agreed to at Cancún."[11] Proceeding on the assumption that Paris would involve a hybrid top-down–bottom-up approach, they took note of various ways in which the pledging system could be strengthened: An agreement could specify what type of actions a state might take to limit emissions, it could bound but not eliminate a state's discretion in stating the ambition of its contribution, or it could—as New Zealand proposed—enumerate general parameters for INDCs while allowing states to opt out of one or more parameters.[12] Other possible elements of flexibility could include modified accounting rules (as Switzerland proposed), a requirement that INDCs be submitted to domestic publics for review (for which there is

precedent in other conservationist and environmental agreements), schedules and deadlines for submission of INDCs and
their review, and provisions for strengthening INDCs to account
for evolving science and capabilities, social, demographic, technological and political developments, natural disasters, and
sundry other "unknown unknowns."

"Treaty mechanisms that allow a party to revise its commitment upward allow parties to lock in stronger measures
internationally," Bodansky and Diringer observed. "For example,
the Ramsar Wetlands Convention allows parties to list additional wetlands unilaterally. Over time, this has led to a huge
increase in the number of wetlands listed as internationally
important under the Ramsar Convention."[13] Conversely, fail-safe
provisions, for which there is precedent in the GATT trade
agreements, could allow for downward revision only in certain
well-specified circumstances.

Japan had proposed formal international review of
INDCs, Bodansky and Diringer noted. Parties to the convention,
NGOs, and private-sector entities might be permitted to submit
questions and comments on the national pledges to Framework
Convention committees, which would consider the INDCs in
formally constituted review sessions.

Not least among the possible elements of flexibility was
the legal form of a Paris agreement. All parties to the climate
negotiations understood all too well that any strong agreement
in treaty form would stand little chance of being ratified by the
US Senate. If the Paris agreement had to take the form of a
treaty, then in effect the Senate would be holding hostage the
whole world's effort to contain the threat of catastrophic climate
change.[14] Fortunately, that problem could be fudged, as Bodansky spelled out here and elsewhere: the Paris agreement could
be an executive agreement that the president reaches strictly on

his or her own authority, an executive agreement with which Congress concurs by regular legislative means, an agreement that merely acknowledges domestic actions already taken, and so on. "The president would be on relatively firm legal ground accepting a new climate agreement with legal force, without submitting it to the Senate or Congress for approval, to the extent it is procedurally oriented, could be implemented on the basis of existing law, and is aimed at implementing or elaborating the UNFCCC [the Framework Convention]."[15] A significant disadvantage to all such procedures, is that any commitments made are more easily reversed by subsequent presidents, Bodansky conceded. But it bears noting that all international agreements are subject to backsliding and the vagaries of domestic politics, whatever form they take.

Although the Europeans had "accomplished a lot," as Jacoby put it, and were going into Paris committed in principle to seeking legally binding emissions cuts, Jacoby and others generally believed that Europe knew that this position was both unrealistic and unachievable.[16] A much more explosive issue was the matter of financial assistance to developing countries. This was a matter of huge importance to the G-77/AOSIS bloc, and legitimately so, as those states worry increasingly about damage limitation, adaptation, and emissions reduction. Jacoby, Bodansky, and others agreed that dangerous climate change would not be containable in the medium and long run unless the poor countries of the world get much more help than they are getting. Yet the industrial countries had been extremely slippery about their Copenhagen commitment to give at least one hundred billion dollars per year to help developing countries cope with climate change.[17] The US Congress flatly refused to countenance any funds being authorized in fulfillment of the promise. If the Paris agreement contained binding language requiring the rich countries to deliver concretely on the promise, that alone could make the agreement dead on arrival in the United States.

Bodansky argued, with considerable support from other experts and players, that the most plausible form of a Paris agreement would be a relatively short binding text, with national pledges and related material somehow attached. The general idea was that parties to the agreement would make a legally binding pledge to take certain actions, which would be specified in the attachments without being binding as such. The very lengthy draft text that emerged from the Lima Conference of Parties in December 2014 bore little resemblance to that vision, however, and the text got only slightly shorter and less convoluted in the negotiations that followed in the next nine months. Bodansky conceded at the beginning of 2015 that a lot of work remained to be done. "The Lima decision kicked many of the substantive issues down the road. The question is how long the parties can continue doing so, if there is to be a successful outcome in Paris."[18]

At the United Nations, Selwyn C. Hart anticipated that the Paris agreement would have four major components: the main text of the agreement itself, the INDCs, an action agenda, and provisions for finance. The Intended Nationally Determined Contributions needed to improve on previous commitments, and backsliding should not be tolerated, he said. With all the world's countries looking to be part of the Lima-Paris process, there needed to be higher levels of ambition and sustenance of momentum. The Framework Convention's Green Climate Fund should be operationalized. The principle of common but differentiated responsibility might be reaffirmed and perhaps reformulated. Limitation of emissions ("mitigation") and adaptation should be put on an equal footing.[19]

In the final months approaching Paris, it was evident to all that if there were an agreement, it would not be one with relatively simple and easily summarized provisions, like the basic principles found in the Rio Convention or the rich-country cuts specified in the Kyoto Protocol. Rather, it would be a complex document with several orders of statements and a great many details, many of them quite important. Full and adequate assessment of the treaty—above all, whether it might get the world heading in the direction of a 2°C path—would not be done hastily, in a night, on little sleep, in alien hotel rooms, by journalists working on tight deadlines.

Epilogue: The Paris Agreement

The two-week global climate conference that ended in Paris on December 12, 2015, the twenty-first of the so-called Conferences of Parties held yearly pursuant to the 1992 UN Framework Convention on Climate Change, could not have been more different from Copenhagen. The French erected a gigantic temporary conference complex in the suburb of Le Bourget, near where Charles Lindbergh landed in 1927. Everybody who was credentialed got in almost instantly on the first try, and we all got through security in a matter of minutes every day thereafter, despite the dreadful terrorist killings that had occurred in Paris just two weeks before. Shuttle buses ran continuously and around the clock from the nearest metro station to and from the conference center. All credentialed participants got all local transportation free of charge, compliments of the French hosts. Helpful guides were all over the place, not just at Le Bourget but in Paris rail and metro stations as well, at the ready to guide participants to their destination. Everybody—delegation members, NGO members, even members of the press—was in a good mood from day one. Everybody sensed that this time something historic really was going to happen.

In Denmark, management of the conference had been uneven from the start, partly because of splits within the government, but partly also—to be brutally frank—because the leaders of that little country seemed way over their heads managing a very large, very complicated international conference. The French knew what they were doing. During the year before the conference, government leaders had traveled the world, conducting bilateral talks in anticipation of every sticky issue that would come up. President François Hollande went to Beijing, for example, and got the Chinese leadership to agree in principle that national climate pledges would be reviewed independently every five years—one of the most important elements, perhaps *the* most important element, in the accords to be negotiated. Foreign Minister Laurent Fabius, as official host of the meeting, took direct personal charge and by all accounts handled every detail beautifully, from the first day when heads of state representing 150 countries convened to the all-night negotiations during the final three days of the meeting.

In the opening leaders' ceremony, featuring the Secretary-General of the United Nations, the French president, the outgoing chairperson of the climate negotiations, and the chairperson of the climate secretariat in Bonn, everybody was reading from the same script. It was as if they all had carefully compared notes before the meeting, and indeed, they almost certainly *had* compared notes.

President Hollande, in his welcoming remarks, asked what would enable us to say that the Paris agreement is good, even "great." First, regular review and assessment of commitments, to get the world on a credible path to keep global warming in the range of 1.5–2°C. Second, solidarity of response, so that no state does nothing and yet none is "left alone." Third, evidence of a comprehensive change in human consciousness, allowing eventually for introduction of much stronger measures, such as a global carbon tax.

Secretary-General Ban Ki-moon articulated similar but somewhat more detailed criteria: the agreement must be lasting, dynamic, respectful of the balance between industrial and developing countries, and enforceable, with critical reviews of pledges even before 2020. He noted that 180 countries had now submitted climate action pledges, an unprecedented achievement, but stressed that those pledges needed to be progressively strengthened over time.

Remarkably, not only the major sponsors of the conference were speaking in essentially the same terms, but civil society was as well. Starting with its first press conference on opening day and at every subsequent one, representatives of the Climate Action Network, representing nine hundred NGOs, confined themselves to making detailed and constructive suggestions about how key provisions of the agreement might be strengthened. Though CAN could not possibly speak for every single one of its member organizations, the mainstream within CAN clearly saw it as the group's goal to obtain the best possible agreement, not to advocate for a radically different treaty. The mainstream would not be taking to the streets.

This was the main item that made Paris different not just from Copenhagen but from every previous climate meeting: Before, there always had been deep philosophical differences between the United States and Europe, between the advanced industrial countries and the developing countries, and between the official diplomats and civil society. At Paris, it was immediately obvious that everybody, nongovernmental organizations included, was "playing from the same book," as I said in a report that first day for fusion.net (a venture at that time of Univision and Disney). So it was evident that an important agreement could be reached and probably would be reached without much dramatic discord. (That may partly account for why 90 percent of the press present on the first day departed, leaving a small cadre of us to follow proceedings in detail.)

In the coming days, national delegations would stake out tough positions, and there would be some hard bargaining. But at every briefing and in every interview, no matter how emphatic the stand, it was made clear that compromises would be made and that nothing would be allowed to stand in the way of agreement.

The two-part agreement formally adopted in Paris on December 12, 2015, represents the culmination of a twenty-five-year process that began with the negotiations in 1990–91 that led to the adoption in 1992 of the Rio Framework Convention. Only with the Paris accords, for the first time, did all the world's nations agree on a common approach that rebalances and redefines respective responsibilities while further specifying what exactly is meant by dangerous climate change. Paragraph 17 of the Decision (or preamble) notes that national pledges will have to be strengthened in the next decades to keep global warming below 2°C or close to 1.5°C, and article 2 of the more legally binding Agreement says that warming should be held "well below 2°C" and if possible limited to 1.5°C. Article 4 of the Agreement calls on those countries whose emissions are still rising to have them peak "as soon as possible," so "as to achieve a balance between anthropogenic emissions by sources and removals by sinks of greenhouse gases in the second half of this century"—a formulation that replaced a reference in article 3 of the next-to-last draft calling for "carbon neutrality" by the second half of the century.

"The wheel of climate action turns slowly, but in Paris it has turned. This deal puts the fossil fuel industry on the wrong side of history," commented Kumi Naidoo, executive director of Greenpeace International.

The Climate Action Network, in which Greenpeace is a leading member, along with organizations like the Union of Concerned Scientists, Friends of the Earth, the World Wildlife Fund, and Oxfam, would have preferred language that flatly adopted the 1.5°C goal and that called for complete "decarbonization"—an end to all reliance on fossil fuels. But to the extent that the network can be said to have common positions, it would be able to live with the Paris formulations, to judge from many statements made by leading members in CAN's twice- or thrice-daily press briefings, as well as statements made by CAN leaders embracing the agreement.

Speaking for scientists, at an event anticipating the final accords, Hans Joachim Schellnhuber, leader of the Potsdam Institute for Climate Impact Research, said with a shrug that the formulation calling for net carbon neutrality by midcentury would be acceptable. His opinion carried more than the usual weight because he sometimes is credited in the German press with being the father of the two-degree standard. Schellnhuber told me in Paris that a Potsdam team had indeed developed the idea of limiting global warming to 2°C in total and 0.2°C per decade; and that while others were working along similar lines, he personally had drawn the Potsdam work to the attention of Angela Merkel in 1994, when she was serving as Germany's environment minister.

As for the tighter 1.5°C standard, this is a complicated issue that the Paris accords fudged a bit. The difference between impacts expected from a 1.5°C world and a 2°C world are not trivial. The odds of a complete Greenland ice sheet melt might be significantly higher in the 2°C scenario than in a 1.5°C world, for example. But at the same time the scientific consensus is that it would be virtually impossible to meet the 1.5°C goal because on top of the 0.8–0.9°C of warming that already has occurred, another half degree is already in the pipeline, "hidden away in the oceans," as Schellnhuber put it. At best, we might be able to work our way back to 1.5°C in the 2030s or 2040s, after first overshooting it. Thus, though organizations like 350.org and scientists like James Hansen continued to insist that 1.5°C degrees should be our objective, pure and simple, the scientific community and the CAN mainstream were reasonably comfortable with the Paris accords' "close as possible" language.

The Paris accord consists of two parts, a long preamble called the Decision and a legally binding part called the Agreement, primarily to satisfy the Obama administration's concerns about having to take anything really sticky to Congress. The general idea, which was developed by the administration with expert legal advice from organizations like the Center for Climate and Energy Solutions, was to put such substantive matters as how much the United States will actually do in coming decades to cut greenhouse gas emissions into the preamble and to confine the treatylike Agreement as much as possible to such procedural issues as when in the future countries will talk about what.

Nevertheless, the distinction between the Decision and the Agreement is far from clear-cut, and the Decision (or preamble) is not short, contrary to some expectations. All the major issues that had to be balanced in the negotiations—not just the 1.5–2.0°C target and the decarbonization language, but financial aid, adaptation and resilience, differentiation between rich and poor countries, reporting requirements, and review—are addressed in both parts. There is nothing unusual as such about an international agreement having two parts, a preamble and a main text. What *is* a little odd about Paris, however, is that the preamble, at nineteen pages, is considerably longer than the eleven-page Agreement, as Beijing-based Chee Yoke Ling of the Third World Network pointed out in Paris. The length

of the Decision, she explained, reflected more than just US
concerns about obtaining Senate ratification. It also arose from
anxieties shared by developing countries about agreeing to
legally binding provisions that might be hard to implement as
well as politically dangerous.

In one of the Paris accord's most important provisions,
the Decision or preamble states that the national pledges are to
be collectively reassessed in a "facilitative dialogue among parties" in 2018, to determine how much progress is being made
toward the agreement's long-term goal. An equally important
provision, found in the Agreement, calls for a global "stocktake"
to be conducted in 2023 and every five years thereafter, covering
all aspects of the agreement's implementation, including its
highly contested provisions about financial aid and "loss and
damage"—the question of support and compensation for
countries and regions that may face extinction as a result of
global warming. Thus, not only carbon-cutting efforts but
obligations of the wealthy countries to the poor will be subject
to the world's critical scrutiny at regular intervals.

The general idea is to exert peer group pressure system-
atically on scheduled occasions, so that everybody will ratchet
up carbon-cutting ambitions. These key requirements for review are very close to those advocated at the conference by the
Climate Action Network and by diplomatic members of the
so-called high ambition group. (The high ambition group was
a loose alliance of developing countries and some industrial
countries, including Canada, which was already taking a radically different diplomatic direction under its newly elected
prime minister, Justin Trudeau.)

On the critical issue of financial aid for developing countries
struggling to reduce emissions and adapt to climate change,

Paris affirms the Copenhagen promise of one hundred billion dollars by 2020, in the Decision (paragraph 115) but not in the more binding Agreement—to the displeasure of the developing countries, no doubt. In the three previous draft versions of the accord, the one-hundred-billion-dollar pledge was contained in the Agreement as well.

Somewhat similarly, the loss and damage language contained in the Decision specifically excludes liability on the part of the advanced industrial countries that have been primarily responsible for climate change until now. This was a disappointment to representatives of the nations and regions most severely and imminently threatened by global warming, but any finding of liability would have been an absolute show-stopper for the American delegation. Still, the fact that loss and damage was even broached represented a victory for the developing world and its advocates, who had been complaining for decades about the complete absence of the subject from the Rio Convention and Kyoto Protocol, despite coverage of loss and damage in many other international environmental agreements.

The so-called G-77 group, which actually represents 134 developing countries plus China, appears to have played a shrewd and tough game at Le Bourget. Its very able and engaging chairperson, South African diplomat Nozipho Joyce Mxakato-Diseko, sent a sharp shot across the prow of the rich countries on the third day of the conference, with a seventeen-point memorandum she e-mailed enumerating the Group of 77's complaints.

"The G77 and China stresses that nothing under the [Framework Convention] can be achieved without the provision of means of implementation to enable developing countries to play their part to address climate change," she said, alluding to the fact that if developing countries are to do more to cut emissions growth, they need help. "However, clarity on the complete picture of the financial arrangements for the enhanced implementation of the Convention keeps on eluding us. . . . We hope that by elevating the importance of the finance discussions under the different bodies, we can ensure that the outcome meets Parties' expectations and delivers what is required."

In a follow-up press conference, Mxakato-Diseko criticized draft language stating that "those in a position to do so" should take responsibility for providing financial assistance. She said the agreement needed to be specific about just who was going to do what. She put it like this: "When I walk into a messy room at home I don't say, 'The room will be cleaned up. I say, Johnnie! Clean up the room!'"

Though the developing countries wanted stronger and more specific financial commitments and loss and damage provisions that would have included legal liability, there is evidence throughout the Paris Decision and Agreement of the industrial countries' giving considerable ground to them. During the formal opening of the conference, President Obama met with leaders of AOSIS—the Alliance of Small Island States—and told them that he understood their concerns as he, too, is "an island boy." (Evidently that went over well.) The reference to the one-hundred-billion-dollar floor for financial aid surely was removed from the Agreement partly because the White House at present cannot get Congress to appropriate money for any climate-related aid. But at least the commitment remained in the Decision, which was not a foregone conclusion.

The one area in which the developing countries *gave* a lot of ground in Paris was in measuring, reporting, and verification (MRV). Under the terms of the Rio Convention and Kyoto Protocol, only the advanced industrial countries—the so-called

Annex 1 countries—were required to report their greenhouse gas emissions to the UN climate secretariat in Bonn. Extensive provisions in the Paris Agreement now call on *all* countries to report emissions, according to standardized procedures that are to be developed.

The climate pledges that almost all countries submitted to the UN in preparation for Paris, the Intended Nationally Determined Contributions, provided a preview of what this will mean. The previous COP, in Lima, had called for all the INDCs to be submitted by the summer and for the climate secretariat to do a net assessment of them by the end of October, which seemed ridiculously late in the game. But when the results of that assessment were released on October 30, the secretariat's head, Christiana Figueres, cited independent estimates that together the INDCs might put the world on a path to 2.7°C warming. That result was a great deal better than most specialists following the procedure would have expected, myself included. Although other estimates suggested the path might be more like 3.5°C, even that figure was a good deal better than the business-as-usual path, which would be at least two degrees higher than that by century's end.

The INDC and the review system are really the foundation stones of the Paris accords. Speaking at a number of Paris side events, V. Ramanathan of the Scripps Institution of Oceanography in San Diego called the INDC a "brilliant invention" and a "fantastic idea." What he always hastened to add was that everything will now depend on how the Paris accords are implemented.

The Paris outcome reflects a sea change in global public opinion inasmuch as educated leadership classes everywhere concluded that climate change is indeed a serious problem that needs to be addressed globally, contrary to what fossil industry lobbyists had been arguing for decades. But the fossil industry is merely down and not altogether out, as Marlowe Hood observed in an assessment of COP 21, "Less than Meets the Eye," published by Agence France Presse.

"What happened in Paris is only a first step, a collective Letter of Intent. . . . Even if powerful vested interests have executed a tactical retreat, they have yet to play their final hand," said Hood, a leader of AFP's twenty-person team at the climate conference. Not only fossil interests, he might have added, but everybody's interest in economic growth and prosperity will vie with the goals of the COP 21 accords.

Among the casualties in the Paris outcome was any talk, for the time being, of global carbon taxation. Six world leaders joined in a formal statement on the first day of the conference advocating a global price on carbon. They must have known that such language would not get into the final agreement, and yet they wanted it to be known that this will be an item on future agendas.

The Clean Development Mechanism (see chapter 3) may have been another casualty. Without the mandatory emissions cuts set out in the Kyoto Protocol, there is no internationally agreed upon cap to provide the basis for internationally trading in emissions permits. Regionally and nationally, emissions trading systems will continue to be built out, of course. But for the time, there will be no talk of linking them up globally under a common rubric. Instead, article 6 of the Agreement calls rather vaguely for the establishment of "a mechanism to contribute to the mitigation of greenhouse gas emissions and support sustainable development." Among other things it is "to contribute to the reduction of emissions levels in the host Party, which will benefit from mitigation activities resulting in emission reductions that can also be used by another Party to fulfill its nationally determined contribution [its INDC]."

Considered as a whole, the Paris outcome will be a success only if there is a sustained self-critical effort by all important parties over several decades. There is some reason to worry whether global publics will stay focused on the issue and continue to bring pressure to bear. But there also is reason to think that bad things will continue to happen as the world warms so that concern and action will be continually reignited.

Only with time, and quite a lot of time at that, will we know whether a system of voluntary commitments made by all states produces more effective results than a system in which selected countries undertake mandatory actions. "Whether the Agreement is truly successful, whether this foundation for progress is effectively exploited over the years ahead by the Parties to the Agreement, is something we will know only ten, twenty, or more years from now," concluded Harvard's Robert Stavins, in a December 12 blogpost about the agreement.

Who deserves credit for Paris, to the extent that credit is due? The United States had played a complicated and at times embarrassing role in climate diplomacy from the beginning. Yet at the 2007 Bali climate conference, which set the agenda for Copenhagen, President George W. Bush's diplomatic representatives suggested a system of voluntary pledging, rather than mandatory emissions cuts. After Copenhagen that idea came to be systematized in the INDC.

Though this approach met with considerable skepticism, the United States had surprising success in the next years in winning allies over to the INDC, with the result that most countries submitted climate action plans, many of them considerably stronger than one might have expected. Part of the credit for that goes to Obama's chief climate negotiator, Todd Stern, who turned on a dime from being the tough guy at Copenhagen doing a nasty but necessary job to being an extremely tactful and modest cajoler. Credit goes too, of course, to Stern's boss Secretary of State John Kerry, who always has been firmly committed to strong climate action, and to Kerry's reputation around the world (not only for that, but also because of his attempt to unseat Bush in 2004 and, just as much, the role he played as a young man as a leading opponent of the Vietnam War).

In a kind of stump climate speech Kerry delivered repeatedly in the months leading up to Paris, he derided those who belittle climate science because they're "not scientists." Oh, so they don't think they know why apples fall off trees or that the world is round, he would ask rhetorically. In the version of the speech he delivered on arrival in Paris, he said that the flat-earthers seem to think that as the world's oceans rise, the water is just going to pour off the sides.

At Paris and before, the Obama administration chose largely to lead from behind, leaving its main European allies— the world leaders in addressing climate change—to carry the ball. At Paris, the American delegation focused on making sure that everybody appreciated just how much the United States is doing, now that it no longer is sitting on its hands. The Environmental Protection Agency's administrator Gina McCarthy made repeated appearances to explain the US clean power plan. California governor Jerry Brown told people what California was doing, and former New York mayor Michael Bloomberg talked about cities (and of course, as usual, himself). Captains of industry, with Bill Gates in the lead, talked about their efforts.

America's faith in its allies proved well-founded. France's Laurent Fabius seemed to enjoy everybody's full confidence, and Germany's diplomats and scientists were ubiquitous, doing everything they could to advance the shared agenda: a treaty that would be binding, enduring, universal, fair, credible, and— above all—effective. True to form, Angela Merkel made only a brief appearance on the first day and never sought the limelight, and yet her fingerprints were everywhere.

In 1994, as *Time* magazine noted when naming her "person of the year," Merkel was host of the first Conference of Parties in Berlin, which set the agenda for Kyoto. That was the year Schellnhuber told her about the logic behind limiting warming to 2°C. In subsequent years, after becoming chancellor, Merkel would sell first Europe and then the whole world on the 2°C goal. At home, she carried on with the aggressive clean-energy policies put in place by the previous socialist-Green government, giving the world its most vivid demonstration yet that economic success and prosperity can coexist with sharp carbon cutting. In the run-up to Paris, the *Guardian* reported, Merkel got Russia's Vladimir Putin to promise his fossil-rich country would not block an effective deal.

The G-77 and China also played on the whole a constructive role. While always seeking to keep the focus on adaptation and resilience funding, the group also made clear that it wanted a strong agreement to be reached. Although Saudi Arabia collected its fair share of "fossil of the day" awards, even it went along in the end with the G-77 consensus. Before Paris, Saudi Arabia submitted a surprisingly strong INDC saying that the country needed to think in terms of diversifying its economy away from oil and gas. Aramco joined a bloc of European oil companies—and parted ways with the American majors—in calling for a more constructive attitude toward climate change.

In the final analysis, however, the main credit must go to the French for making Paris such a definite success, and in this there is a nice symmetry. For it was a Frenchman, Jean Ripert, whose brilliant diplomatic work in 1991–92 as chairman of the United Nations' International Negotiating Committee on Climate Change established the basis for the UN Framework Convention on Climate Change. France's Fabius, as manager of the almost fantastically complex Paris negotiations, helped forge a path that may actually make the Framework Convention work.

Appendix 1: Survey, Climate Summit Marchers, September 21, 2014 ($N = 100$)

In marching here today, do you wish to bring pressure on all nations and leaders to address climate change more aggressively?

yes 98%   no 2%

Is coordinated international action essential to addressing climate change effectively?

yes 94%   no 6%

Should the United States take stronger action on climate change regardless of whether other countries go along?

yes 100%   no

Which of these countries do you see as impediments to strong international action on climate change?

Brazil	China	Germany	India	United States	United Kingdom
19%	75%	12%	42%	80%	21%

Are you hopeful that a strong, binding climate agreement will be reached by the time of the Paris conference in November 2015?

yes 51%   no 31%   no opinion 18%

Do you approve of the negotiating position the United States has adopted in climate negotiations?

yes 4%   no 65%   no opinion 31%

Reporter John H. Cushman Jr., writing in *Inside Climate News,* has asked whether "these Climate Week events [are] the makings of a turning point in the world's effort to escape the risks of climate change, or a formula for futility?" Do you think today is:

a turning   a formula for   neither 32%
point 60%   futility 2%

Appendix 2: Chronology of Climate Negotiations

1985   Twenty nations, most of them major producers of chlorofluorocarbons (CFCs), adopt the Vienna Convention, which calls for negotiation of a global agreement to regulate and reduce ozone-depleting substances

1987   The Montreal Protocol calls for the complete phase-out of the chemical compounds known to be depleting the earth's ozone shield; the treaty wins near-universal adherence

1988   The Intergovernmental Panel on Climate Change (IPCC) is created by the World Meteorological Organization (WMO) and UN Environment Programme (UNEP); UN General Assembly Resolution 45/53 calls on the WMO and UNEP, via the IPCC, to consider formulation of a global climate convention

1990   UN General Assembly Resolution 45/212 establishes an Intergovernmental Negotiating Committee (INC) to formulate a global climate convention

1992   In the context of an Earth Summit in Rio, world leaders adopt the United Nations Framework Convention

on Climate Change (UNFCCC), which establishes the
basic principles and procedures that will govern future
climate diplomacy; the treaty promptly wins ratification by virtually all nations

1995   The IPCC issues guidelines for national reporting
of greenhouse gas inventories, in anticipation of negotiated emission cuts that will be required of advanced
industrial countries, pursuant to the Rio Framework
Convention

1995   The first Conference of Parties (COP 1), meeting in
Berlin pursuant to the Rio Convention, adopts a mandate calling for negotiation of specific cuts to be made
in greenhouse gas emissions by the advanced industrial
countries

1997   The Kyoto Protocol, adopted at COP 3, specifies cuts in
greenhouse gas emissions that the advanced industrial
countries will be required to make in six greenhouse
gases by 2008–12 from 1990 levels; the total combined
cut made by specified industrial countries is to be
5 percent. The protocol also gives its blessing to market
devices as means of cutting emissions and establishes
the Clean Development Mechanism (CDM), so that
parties in developing countries can sell offsets to counterparts in industrial countries. A carve-out for the
economically depressed former Soviet states allows
them to sell emissions offsets (dubbed "hot air" by
critics among nongovernmental organizations)

1998   BP, Monsanto, GE, and the World Resources Institute
publish *Safe Climate, Sound Business,* which lays the
basis for a greenhouse gas protocol that will be widely
used by private-sector and nongovernmental organizations tracking or trading greenhouse gas emissions

2001   COP 7 adopts the Marrakech Accords, which include guidelines for operation of the CDM, a revised listing of countries eligible for green development funding, and the creation of an Adaptation Fund

2005   The Kyoto Protocol takes legal force, following its ratification by Russia the previous year

2007   COP 13 in Bali, Indonesia, adopts a mandate calling for agreement by end-2009 on a "long-term global goal for emissions reduction" and on "enhanced action" regarding greenhouse gas reduction ("mitigation"), climate change adaptation, technology transfer, international cooperation, and financing

2007   The IPCC Fourth Assessment Report finds that human activity is unequivocally responsible for climate change

2007   Former US vice president and senator Al Gore is awarded the Nobel Peace Prize, together with the IPCC volunteer scientists

2009   COP 15 in Copenhagen, Denmark, widely expected to produce a new legally binding treaty that would build on the Kyoto Protocol and establish a new program of mandatory emissions cuts for the period 2012–20, instead agrees informally on holding global warming to 2 degrees Celsius and on initiating a program of voluntary emissions goals by all major emitting countries

2010   COP 16 in Cancún, Mexico, confirms the 2°C goal and creates framework for the Green Climate Fund, to assist less-developed countries with greenhouse gas reduction and climate adaptation

2011   COP 17, meeting in Durban, South Africa, establishes a working group (the so-called ADP, or Ad Hoc Working Group on the Durban Platform) to "develop a [climate] protocol, another legal instrument or an agreed outcome with legal force under the [Rio Framework] Convention applicable to all Parties, which is to be completed no later than 2015"

2013   COP 19 establishes the Warsaw Mechanism for loss and damages associated with climate change, to be managed by an executive committee reporting to the Conference of Parties. Its rather vague mandate is, among other things, to develop understanding of risk management, strengthen dialogue and cooperation by major stake-holders, and encourage "action and support, including finance, technology and capacity-building, to address loss and damage"

2013/14 COPs 19 and 20 in Warsaw, Poland, and Lima, Peru, formalize a system of national climate pledges to be submitted to the climate secretariat in Bonn, the Intended Nationally Determined Contributions (INDCs)

2014   The IPCC Fifth Assessment Report predicts that the average global temperature will increase by 4.8°C and that oceans will rise 1 meter in this century on business-as-usual projections

2015   COP 21 meets in Paris with the objective of adopting a new comprehensive, long-term climate agreement, in keeping with the Durban Platform

Notes

1
Can Catastrophic Climate Change Be Averted?

1. Richard Kinley, interview with the author, Apr. 14, 2014.

2. Carlarne, *Climate Change Law and Policy,* 254; Gupta, *History of Global Climate Governance,* 7.

3. Nordhaus, *Climate Casino,* 198.

4. Justin Gillis, "3.6 Degrees of Uncertainty," *New York Times,* Dec. 15, 2014.

5. For background, see *Wikipedia,* s.v. "Avoiding Dangerous Climate Change," https://en.wikipedia.org/wiki/Avoiding_dangerous_climate_change. According to a Hadley Centre ensemble of climate models, there would be a 78% chance of 2°C warming at 450 ppm and a 99% chance at 550 ppm: Posner and Weisbach, *Climate Change Justice,* 18, table 1.1. For a deeper account of scientific thinking on the eve of the Copenhagen conference, see Malte Meinshausen et al., "Greenhouse-Gas Emission Targets for Limiting Global Warming to 2°C," *Nature* 458 (Apr. 30, 2009): 1158–62; and Richard Monastersky, "A Burden beyond Bearing," *Nature* 458 (Apr. 30, 2009): 1091–94. See also International Symposium on the Stabilisation of Greenhouse Gases, Hadley Centre, Met Office, Exeter, Report of the Steering Committee, Feb. 3, 2005, http://www.g8.utoronto.ca/environment/2005steeringcommittee.pdf.

6. The IPCC estimate was part of its Fifth Assessment Report, one of the regular reports the IPCC issues under the auspices of the UN Environment Programme and the World Meteorological Organization. For a parallel account, see Myles R. Allen and Thomas Stocker, "Warming Caused by Cumulative Carbon Emissions towards the Trillionth Tonne," *Nature* 458 (April 30,

2009): 1163–66. For context, see Fred Pearce, "The Trillion Ton Cap: Allocating the World's Carbon Emissions," *Yale Environment 360,* Oct. 24, 2013, http:// e360.yale.edu/feature/the_trillion-ton_cap_allocating_the_worlds_carbon_ emissions/2703/; and Christophe McGlade and Paul Ekins, "The Geographical Distribution of Fossil Fuels Unused When Limiting Global Warming to 2°C," *Nature* 517 (Jan. 8, 2015): 187–90.

7. McKibben in turn was taking his cues from James Hansen, the immensely influential modeler. (For background on Hansen and his significance, see Sweet, *Kicking the Carbon Habit,* chaps. 5, 6.) Hansen's optimism about the achievability of 350 ppm seems to rest in part on a relative optimism concerning the quantities of greenhouse gas we can still afford to emit, consistent with that goal: see Fred Pearce, "What Is the Carbon Limit? That Depends Who You Ask," *Yale Environment 360,* Nov. 6, 2014, http://e360.yale. edu/feature/what_is_the_carbon_limit_that_depends_who_you_ask/2825/.

8. David G. Victor and Charles F. Kennel, "Climate Policy: Ditch the 2°C Warming Goal," *Nature* 514 (Oct. 2, 2014): 30–31.

9. Quoted in Alex Morales, "Kyoto Veterans Say Global Warming Goal Slipping Away," *Bloomberg,* Nov. 4, 2013, http://www.bloomberg.com/news/ articles/2013-11-04/kyoto-veterans-say-global-warming-goal-slipping-away.

10. Kiley Kroh, "Germany Sets New Record, Generating 74 Percent of Power Needs from Renewable Energy," May 13, 2014, http://thinkprogress. org/climate/2014/05/13/3436923/germany-energy-records/; Mark Hertsgaard, *Nation,* Apr. 6, 2016, http://www.thenation.com/article/203601/3-reasons-be-optimistic-about-fight-save-climate/.

11. Marie Charrel, "La pensée verte des pays nordiques," *Le Monde,* Oct. 4–5, 2015.

12. William Fehrman, president and CEO, MidAmerican Energy Company, Drake University Forum, "The Frontier of Climate Change," panel co-sponsored by the League of Women Voters and the *New Republic,* Apr. 22, 2014.

13. World Bank, "Turn Down the Heat: Confronting the New Climate Normal," Report No. 3, November 2014, http://www.worldbank.org/en/topic/ climatechange/publication/turn-down-the-heat, executive summary, p. 2. The UNEP *Emissions Gap Report 2014* was prepared at the behest of the German parliament. When a preliminary interim version of Sachs et al., *Pathways to Deep Decarbonization,* was issued in mid-2014, Columbia University went to some lengths to publicize it. (I took part in two press briefings, on Dec. 3, 2013, and July 7, 2014.) But when the final version appeared a year later, neither Columbia nor the United Nations called it to the attention of the press, to my knowledge.

14. All numbers for per capita emissions are from the World Bank and represent yearly averages for the period 2010–14.

15. Sachs, *Age of Sustainable Development,* 201 (fig. 6.10), 397.

16. Press briefing, *Pathways to Deep Decarbonization* report, Columbia University, July 7, 2014.

17. Stephen Cass, a specialist on space technology at *IEEE Spectrum,* enumerates the following for John F. Kennedy's moon program: heavy-lift rockets, space rendezvous and docking, computerized guidance and control technology, and (not least) space suits, which would be "basically portable spacecraft." As *Spectrum*'s former specialist on nuclear technology, I would make a list that would include uranium enrichment and plutonium production (reactor) technology, as well as what now goes by the name of "weaponization"—especially the configuration of implosion casings and implosion timing circuitry.

18. Pilita Clark, "ArcelorMittal to Turn Toxic Waste into Biofuel," *Financial Times,* July 13, 2015.

19. Victor, *Climate Change,* 94.

20. "In 2005, when many of the major investments in unconventional oil were getting under way, the EIA [US Energy Information Administration] projected that global oil demand would reach 103.2 million barrels per days in 2015; now it's lowered that figure for this year [2015] to only 93.1 million barrels"; Michael Klare, "Is Big Oil Finally Entering a Climate Change World?," *TomDispatch.com,* Mar. 14, 2015, http://www.tomdispatch.com/blog/175967/tomgram%3A_michael_klare,_is_big_oil_finally_entering_a_climate_change_world/.

21. A crucial consideration is this: If a coal-fired power plant is replaced by a gas plant, the immediate effect is to cut greenhouse gas emissions in half; but once the gas plant is installed, carbon dioxide will be emitted for decades—fossil fuel emissions are locked in, so to speak. For an influential analysis, see David and Socolow, "Commitment Accounting."

22. According to the highly regarded atmospheric chemist Susan Solomon, there is a threefold uncertainty, and that range has not narrowed in thirty years of research; Pearce, "What Is the Carbon Limit?" Bear in mind, though, that the earliest computer estimates of how much warming would result from a doubling of greenhouse gases—roughly 1.5–4.5°C—have scarcely budged in fifty years. Hansen, who created one of the first models to produce that estimate, says today that paleoclimatology gives us a more certain estimate than modeling can provide: press briefing, *Pathways to Deep Decarbonization* report, Columbia University, Dec. 3, 2013.

2

What Else Is at Stake?

1. Joe Romm, "2ºC or Not 2ºC: Why We Must Not Ditch Scientific Reality in Climate Policy," *Climate Progress*, Oct. 1, 2014. http://thinkprogress. org/climate/2014/10/01/3574471/2c-climate-policy/.

2. Joseph Alcamo wrote the Introduction with Daniel Puig and Joeri Rogelj; UNEP, *Emissions Gap Report 2014*, 1.

3. Nordhaus, *Climate Casino*, 177–78. That finding is broadly consistent with the 2006 *Stern Review on the Economics of Climate Change*, which is readily accessible online.

4. Nordhaus, *Climate Casino*, 179.

5. Rahmstorf and Schellnhuber, *Klimawandel*, 94–98.

6. In standard cost-benefit analysis, future costs and benefits are discounted to reflect the fact that we care more about what is happening to us right now than what will happen to us or others in the future. But when the technique is applied to very long-range issues like climate change, in the medium term estimated costs and benefits tend to cancel, whereas in the long run—the scale of a half-century or a century, say—the discounted value of all costs and benefits tends to zero. As a result, saving the world is assigned no value.

7. Klein, *This Changes Everything*. See Naomi Oreskes and Erik M. Conway, *Collapse of Western Civilization* and *Merchants of Doubt*. Pope Francis's papal encyclical *Laudato Si* also is to some extent representative of this school.

8. On rejecting capitalism, Peter Newell and Matthew Paterson put it like this: though they are "highly skeptical of the idea that capitalism can deliver either a socially just or sustainable future," as concerned citizens and activists they "want to see urgent action within short time-frames. This means that post-capitalist futures, while in many ways very attractive, will not provide the political and social context within which we have to tackle this most pressing of issues." Newell and Paterson, *Climate Change*, 14.

9. The only important exception has been Cuba, and today's Cuban leadership appears to have regrets about that.

10. Klein, *This Changes Everything*, 45.

11. Cass Sunstein has been criticized for minimizing the role of both direct regulation and bargaining in protection of the environment and public health. See Robert Kuttner, "Obama's Obama: The Contradictions of Cass Sunstein," *Harper's Magazine*, December 2014, 87–92; and Michael Walzer, "Is the Right Choice a Good Bargain?" *New York Review of Books*, Mar. 5, 2015.

12. Among them Joyeeta Gupta: it was "regrettable that the no-harm, the polluter pays and/or the liability principles [of the Rio Declaration] were not included in the [Framework] Convention." Gupta, *Global Climate Governance,* 93.

13. Speaking at the New School, New York City, May 2015.

14. James K. Boyce, "Amid Climate Change, What's More Important: Protecting Money or People?," *Los Angeles Times,* Dec. 21, 2014.

15. A good place to start is the collection edited by Gardiner et al., *Climate Ethics: Essential Readings.*

16. Stephen M. Gardiner, "A Perfect Moral Storm: Climate Change, Intergenerational Ethics and the Problem of Moral Corruption," in Gardiner et al., *Climate Ethics,* 87–97.

17. Page, *Climate Change, Justice and Future Generations,* 132.

18. Posner and Weisbach, *Climate Change Justice,* 8.

19. Peter Singer, "Changing Values for a Just and Sustainable World," in Held, Fane-Hervey, and Theros, *Governance of Climate Change,* chap. 8. Singer's approach, though attractive, will strike some readers as too simple, almost simple-minded. There are much more sophisticated versions based on the same concept, and some are quite interesting and well developed, but the problem with them is that they tend to get too complex, probably too complicated, to be negotiated. See Paul Baer et al., "Greenhouse Development Rights: A Framework for Climate Protection That Is 'More Fair' Than Equal Per Capita Emissions Rights," in Gardiner et al., *Climate Ethics,* 215–30.

20. See, e.g., Alice Bows and Kevin Anderson, "Contraction and Convergence: An Assessment of the *CCOptions* Model," *Climatic Change* 91 (2008): 275–90.

21. See Simon Caney, "Climate Change, Human Rights and Moral Thresholds," in Gardiner et al., *Climate Ethics,* 171.

22. Siobhan McInerney-Lankford, Mac Darrow, and Lavanya Rajamani, *Human Rights and Climate Change: A Review of the International Legal Dimensions,* World Bank eLibrary, March 2011, http://dx.doi.org/10.1596/978-0-8213-8720-7, 9.

23. John Schwartz, "Ruling Says Netherlands Must Reduce Emissions," *New York Times,* June 25, 2015.

24. Posner and Weisbach, *Climate Change Justice,* 74.

25. See John Gray, "How and How Not to Be Good," *New York Review of Books,* May 21, 2015.

26. Gray, "How and How Not to Be Good." A serious systemic flaw in Posner and Weisbach, *Climate Change Justice,* is that they repeatedly argue against the straw-man notion that it is an objective of the climate regime to redistribute global income. The point of the regime, as embodied in the

principle of common but differentiated responsibility (discussed below in the text), is not to redistribute income as such but to tackle dangerous global warming in a way that is fair and practical.

27. Rawls, *Theory of Justice*, 14.

28. Rawls, *Theory of Justice*, 15.

29. Rajamani, "Differentiation in a 2015 Climate Agreement."

30. Surya Sethi, interview with the author, July 8, 2014.

31. Selwin C. Hart, "The Road to Paris and Beyond: How to Create a Climate Agreement That Works," lecture, The New School, New York, May 11, 2015.

32. Gupta, *History of Global Climate Governance*, 207.

3
Can Diplomacy Deliver?

1. Nordhaus, "New Solution."

2. Timothy E. Wirth and Thomas A. Daschle, "A Blueprint to End Paralysis over Global Action on Climate," *Yale Environment 360*, May 19, 2014, http://e360.yale.edu/feature/a_blueprint_to_end_paralysis_over_global_action_on_climate/2766/.

3. Victor, *Global Warming Gridlock*, 32.

4. Robert J. Shiller, "Putting Idealism to Work on Climate Change," *New York Times*, March 29, 2015.

5. Eduardo Porter, "Rethinking How to Split the Costs of Carbon," *New York Times*, Dec. 25, 2013.

6. Thomas C. Esty, "Bottom-Up Climate Fix," *New York Times*, Sept. 22, 2014.

7. Fred Pearce, "Beyond Treaties: A New Way of Framing Global Climate Action," *Yale Environment 360*, Sept. 29, 2014, http://e360.yale.edu/feature/beyond_treaties_a_new_way_of_framing_global_climate_action/2809/.

8. Klein, *This Changes Everything*, 18.

9. Depledge, "Opposite of Learning," 1.

10. Hoffman, *Climate Governance at the Crossroads*, 6, 16.

11. Gupta, *History of Global Climate Governance*, 171.

12. Cinnamon Carlarne, interview with the author, Dec. 15, 2014.

13. See Gerrard, *Global Climate Change and U.S. Law*, 40.

14. See José Goldemberg, "The Road to Rio," in Mintzer and Leonard, *Negotiating Climate Change*, 177–79.

15. For a good account, see Shapiro, *Carbon Shock*, 73.

16. Shapiro, *Carbon Shock*, 73.

17. On the UN-REDD Programme, see http://www.un-redd.org/aboutredd.

18. Gupta, *History of Global Climate Governance*, 168.

19. Shapiro, *Carbon Shock*, 71.

20. Jeffrey Hayward, interview with the author, July 21, 2015. The Rainforest Alliance had been formed independently of the climate negotiations, before they began, with the objective of saving forests from the intrusion of cattle farming and of paper, pulp, and soy plantations. It played an important role in lobbying climate negotiators to adopt REDD and in helping work the kinks out of the REDD system.

21. Nicholas St. Fleur, "Study Finds Major Threat to Amazon Tree Diversity," *International New York Times*, Nov. 21–22, 2015.

22. See Kitty Stapp, "Small Victories at Bonn Climate Talks," Inter Press Service, June 11, 2015.

23. See Sachs, *Age of Sustainable Development*, 205–6, 339–43. Agriculture is a major source of the three most important greenhouse gases, CO_2, CH_4, and N_2O. It is the top source of nonenergy greenhouse gas emissions, which in turn represent about a third of total emissions.

24. William Sweet, "Food and Climate," in *Great Decisions*, 2014 ed. (Foreign Policy Association, 2013), 68.

25. See Gupta and Van der Grijp, *Mainstreaming Climate Change in Development Cooperation*, 89. Such efforts often do not go much beyond "integration" and tend to be limited to win-win situations, the editors say.

26. Gupta, *History of Global Climate Governance*, 113.

27. Hoffman, *Climate Governance at the Crossroads*, 124.

28. California's trading system applies to utilities, larger businesses, and wholesale gasoline suppliers, whereas the much more limited RGGI applies only to utilities. See Alejandro Lazo, "How Cap-and-Trade Is Working in California," *Wall Street Journal*, Sept. 28, 2014.

29. Richard Sandor, interview with the author, Aug. 12, 2014.

30. Greg Breining, "The Father of Cap and Trade," *Reach*, Winter 2012 (University of Minnesota, College of Liberal Arts).

31. See Sandor, *Good Derivatives*, chaps. 19, 20, esp. 403–10. Sandor's book has been translated into Chinese with a title that translates back into English as "Derivatives Are Not a Bad Child."

32. The most notable loophole involved HFC-23, a waste gas from productions of refrigerants. For background, see Klein, *This Changes Everything*, 231–32; and Shapiro, *Carbon Shock*, 163–64. The episode was covered extensively in the press, and in mid-2013 the European Commission banned traffic in the HFC-23 offsets.

33. L. Hunter Lovins and Boyd Cohen, though often quite critical of market mechanisms, provide a generally positive account of the CDM and its offsets procedures; Lovins and Cohen, *Climate Capitalism*, 233–36.

34. On the CDM as "moribund," see Fiona Harvey, "What Has Changed since Climate Talks in Copenhagen?" *Climate Central*, Aug. 1, 2015, http://www.climatecentral.org/news/what-has-changed-since-copenhagen-19292.

35. Gupta, *History of Global Climate Governance*, 118. The complete up-to-date CDM registry is at http://cdm.unfccc.int.

36. Green, *Rethinking Private Authority*, 122.

37. IPCC, *Fifth Assessment Report*, Working Group 3, chap. 13, p. 63.

38. The origins and workings of this system are described in detail in Green, *Rethinking Private Authority*, chaps. 4, 5.

39. As recently as 2013 some 130 environmental organizations called for scrapping the ETS; Klein, *This Changes Everything*, 231–33, 238–39. But see Shapiro, *Carbon Shock*, 160.

40. Robert Stavins, interview with the author, Sept. 29, 2014.

41. See, e.g., Krugman's review of Nordhaus's *Climate Casino* in *New York Review of Books*, Nov. 7, 2013; and Henry M. Paulson Jr., "The Coming Climate Crash," *New York Times*, June 21, 2014. Sachs pitched the idea of global carbon taxation in a press briefing on the emissions pathways study he spearheaded: press briefing, *Pathways to Deep Decarbonization* report, Columbia University, Dec. 3, 2013.

42. The work of Ian Parry and Chandara Veung, with Dirk Heine of the University of Bologna, was cited by Eduardo Porter, "The Benefits of Curbing Carbon Emissions," *New York Times*, Sept. 23, 2014.

43. So he told me during the Paris climate conference, Dec. 1, 2015,

44. Daniel Bodansky, interview with the author, Sept. 12, 2014.

45. See Institute for Energy Research, "IEA Review Shows Many Developing Countries Subsidize Fossil Fuel Consumption," Nov. 23, 2013, http://instituteforenergyresearch.org/analysis/iea-review-shows-many-developing-countries-subsidize-fossil-fuel-consumption-creating-artificially-lower-prices/; Brandon Baker, "World's Richest Countries Spent $500 Billion on Fossil Fuel Subsidies," *EcoWatch*, Nov. 7, 2013, http://ecowatch.com/2013/11/07/worlds-richest-countries-spent-500-billion-fossil-fuel-subsidies/; and "Eleven Countries with Large Fossil Fuel Subsidies," *National Geographic*, June 20, 2012. The eleven are Iran, Saudi Arabia, Russia, India, China, Egypt, Venezuela, the United Arab Emirates, Indonesia, Uzbekistan, and (last) the United States. *National Geographic* estimated US subsidies at about $15 billion; Oil Change International has put them at $37.5 billion.

46. Merrill et al., *Tackling Fossil Fuel Subsidies and Climate Change*, 9.

47. See, e.g., Sean Sweeney, "Green Capitalism Won't Work," *New Labor Forum* 24, no. 2 (2015): 3.

48. Purvis and Stevenson, *Rethinking Climate Diplomacy*, 24–25. Estimates of so-called after-tax subsidies in the US are much higher, taking account of externalities, the social costs of burning fossil fuels. For a critical discussion, see Aldy, "Policy Surveillance in the G-20 Fossil Fuel Subsidies Agreement."

49. Lovins and Cohen, *Climate Capitalism*, 278–79.

50. Martin Khor, "Cancun Meeting Used WTO-Type Methods to Reach Outcome," *SUNS*, Dec. 16, 2010.

51. Geoffrey Heal, "What Does It Take to Reach a Climate Agreement?" lecture, The New School, New York, Nov. 20, 2014.

52. See William Sweet, "Cell Phones Answer Internet's Call," *IEEE Spectrum*, August 2000, 42–46.

53. Depledge, "Opposite of Learning," 3.

54. Quoted in Hoffman, *Climate Governance at the Crossroads*, 161.

55. Barry Blechman and Ruth Greenspan Bell, "Time to Look beyond the UN Climate Negotiations," *Bulletin of the Atomic Scientists*, Feb. 9, 2014.

56. Nordhaus, "Climate Clubs."

57. Nordhaus, "Climate Clubs," 1367.

58. So far that initiative has met, admittedly, with mixed results, as Aldy, "Policy Surveillance in the G-20 Fossil Fuel Subsidies Agreement," explained. The term "inefficient subsidy" is a big loophole. So is a provision making exception for the protection of low-income fossil energy consumers. The definition of subsidy, as noted earlier, is slippery and hard to pin down. The G-20 countries with by far the biggest fossil subsidies for consumers— Russia, China, and Saudi Arabia—flatly deny that such subsidies exist. Other countries with big subsidies like Iran and Venezuela are not G-20 members and therefore not party to the agreement.

59. See William Sweet, "Clean Air, Murky Precedent," *New York Times*, Sept. 29, 2006.

60. Depledge, "Opposite of Learning," 18.

4

The Superpowers

1. I am grateful to M. V. Ramana of Princeton University for drawing my attention to this important point. Interview with the author, Apr. 30, 2014.

2. To access greenhouse gas emissions data by country, go to the UN Framework Convention on Climate Change site at http://unfccc.int/2860.

php, and then scroll to Process, GHG Data, GHG Data-UNFCCC, Detailed Data by Party.

3. The target referred to an average of the five-year period.

4. See "Germany's Energy Poverty: How Electricity Became a Luxury Good," *Spiegel Online International,* Sept. 4, 2013, http://www.spiegel.de/international/germany/high-costs-and-errors-of-german-transition-to-renewable-energy-a-920288.html.

5. See Peter Sopher, "Germany Is Revolutionizing How We Use Energy . . . and the U.S. Could Learn a Thing or Two," Environmental Defense Fund, May 14, 2014, http://blogs.edf.org/energyexchange/2014/05/14/germany-is-revolutionizing-how-we-use-energyand-the-u-s-could-learn-a-thing-or-two/; and Sopher, "Germany's Energiewende Requires Sophisticated Governance, Political Stamina," Environmental Defense Fund, Dec. 29, 2014, http://blogs.edf.org/energyexchange/2014/12/29/germanys-energiewende-requires-sophisticated-governance-political-stamina/. The Socialist-Green coalition had opted for a nuclear phase-out, but during her first terms Merkel quietly sought "an exit from the nuclear exit" (*einen Ausstieg von dem Ausstieg*), as the German press liked to put it. After Fukushima, she threw in the towel.

6. See Committee on Climate Change, "The Climate Change Act and UK Regulations," https://www.theccc.org.uk/tackling-climate-change/the-legal-landscape/global-action-on-climate-change/.

7. See, e.g., European Commission, Standard Eurobarometer 82, "Public Opinion in the European Union," report, Autumn 2014, available at http://ec.europa.eu/public_opinion/archives/eb/eb82/eb82_en.htm; and Council on Foreign Relations, *Public Opinion on Global Issues,* chap. 5a, World Opinion on the Environment, Nov. 30, 2011, available at www.cfr.org/public_opinion.

8. Purvis and Stevenson, *Rethinking Climate Diplomacy,* 8.

9. It bears noting that during the 1980s and 1990s, when concern about climate change was beginning to spread among experts and governing elites, a nightmarish scenario that got a lot of attention was one in which the Gulf Stream would be disrupted, plunging northwestern Europe into Arctic cold; see Wallace S. Broecker, "The Biggest Chill," *Natural History,* October 1987, 74–81. The common wisdom that the Gulf Stream accounts for northwest Europe's benign climate has not gone uncontested: see Richard Seager et al., "Is the Gulf Stream Responsible for Europe's Mild Winters?" *Quarterly Journal of the Meteorological Society* 128 (October 2002): 2563–85. But no matter how you look at it, Europe's relative mildness is an anomaly.

10. See Alfred Grosser, *The Western Alliance: European-American Relations since 1945* (Vintage Books, 1982); and Tony Judt, *Postwar: A History of Europe since 1945* (Penguin, 2005).

11. Carlarne, *Climate Change Law and Policy,* 275.

12. Hoffman, *Ozone Depletion and Climate Change,* 115. See also Fehl, *Living with a Reluctant Hegemon,* 129–30.

13. Carlarne, *Climate Change Law and Policy,* 273.

14. Carlarne, *Climate Change Law and Policy,* 275.

15. See Gupta and Grubb, *Climate Change and European Leadership,* 81; and Carlarne, *Climate Change Law and Policy,* 251.

16. "Sind Wir Noch zu Retten?" (Are we still to be saved?), *Der Spiegel,* Feb. 21, 2015, 57–64.

17. Simon Schunz, "Beyond Leadership by Example: Towards a Flexible European Union Foreign Climate Policy," Working Paper FG8, January 2011, Global Issues Division, German Institute for International and Security Affairs, 2011.

18. Bauer, "It's about Development, Stupid!," 112–13.

19. Victor, *Climate Change,* 98.

20. The resolution and related documents are reproduced in full as an appendix to Victor, *Climate Change.*

21. See Carlarne, *Climate Change Law and Policy,* 244.

22. The statement, which was shown at a meeting of climate activists in California, reportedly moved some to tears.

23. Betsill, "Environmental NGOs and the Kyoto Protocol Negotiations 1995 to 1997," in Betsill and Corell, *NGO Diplomacy,* 50–51; Gupta, *History of Global Climate Governance,* 98.

24. Harland L. Watson, testimony to the House Foreign Affairs Subcommittee on Asia, the Pacific, and the Global Environment, July 11, 2007: US Government Printing Office, Washington, DC, 2008.

25. Surya Sethi, interview with the author, July 8, 2014.

26. Robert Stavins, interview with the author, Sept. 29, 2014.

27. Betsill, "Environmental NGOs and the Kyoto Protocol Negotiations 1995 to 1997," 50–51.

28. At a meeting of the Carbon Technology Management Conference in Arlington, Virginia, in October 2013, where I delivered a luncheon address, two keynote speakers said that the United States had not only met but exceeded its Kyoto obligations.

29. See Shapiro, *Carbon Shock,* 7–22, for a scathing account of this unedifying episode.

30. White House, Office of the Press Secretary, "Remarks by the President on Climate Change," Georgetown University, Washington, DC, June 25, 2013.

31. Sethi, interview with the author, July 8, 2014. In a talk at Columbia University on Aug. 26, 2013, Energy Secretary Ernest J. Moniz said that the

United States was about halfway to its 2020 greenhouse gas reduction goal of 17% and that about half of that was because natural gas was being substituted for coal in electricity generation.

32. *Vanity Fair*'s special green issue appeared in May 2006; the same month *Wired* had a photograph of Al Gore on the cover, with a feature about "Climate Crisis." *Time*'s famous cover appeared on April 3, 2006.

33. See especially the Yale University–George Mason University poll of October 2014, "Climate Change in the American Mind."

34. Notably, some polls were showing significant numbers of Republican voters expressing concern about global warming. See Suzanne Goldenberg, "Majority of Red-State Americans Believe Climate Change Is Real, Study Shows," *Guardian,* Nov. 13, 2013.

35. White House, Office of the Press Secretary, "Remarks by the President on Climate Change."

36. If Merkel was disgruntled, prominent reasons would have included the discovery that the US intelligence had tapped into her personal cellphone. But she may also have been feeling miffed that Germany got so little credit for its global leadership in green energy adoption and in European climate diplomacy.

37. Purvis and Stevenson, *Rethinking Climate Diplomacy,* 29. Like Daniel Bodansky, Purvis was a high-level climate diplomat straddling the administrations of Bill Clinton and George W. Bush.

5
BRICs, BASICs, and Beyond

1. The BRICs, including now South Africa, have met annually under the chairmanship of Russia. See Brendan O'Boyle, "Explainer: Which Are the BRICs?" July 11, 2014, AS/COA, http://www.as-coa.org/articles/explainer-what-are-brics.

2. Gupta, *History of Global Climate Governance,* 69.

3. For a nice graphical treatment of the coal basics, see Brian Kahn, "What You Need to Know about U.S.-China Climate Pact," *Climate Central,* Nov. 12, 2014, http://www.climatecentral.org/news/details-behind-us-china-climate-pact-18317.

4. For a graphic treatment, see Alexandre Pouchard, "Emissions de CO_2: Ce que pèsent les États-Unis et la Chine," *Le Monde,* Dec. 11, 2014.

5. See Damian Ma, "China's Coming Decade of Natural Gas," in *Energy Security and the Asia-Pacific,* ed. Mikkal E. Herberg (National Bureau

of Asian Research, 2014), chap. 8, and Charlie Zhu, "Drilling Furiously: Chinese Energy Giants Turn Upbeat on Shale Gas," Thomson Reuters, Aug. 29, 2014.

6. See David Sandalow et al., "Meeting China's Shale Gas Goals," working draft, Columbia/SIPA, September 2014; Jaeah Lee and James West, "America's Fracking Boom Comes to China," *Atlantic,* September 2014; "Shale Game: China Drastically Reduces Its Ambitions to Be a Big Shale-Gas Producer," *Economist,* Aug. 30, 2014; and Keith Bradsher, "No Easy Path to Natural Gas," *New York Times,* Aug. 22, 2014.

7. Chi-Jen Yang and Robert B. Jackson, "China's Synthetic Natural Gas Revolution," *Nature Climate Change* 3 (October 2013): 852. See also Simon Denyer, "In China, a Tug of War over Coal Gas: Cleaner Air but Worse for the Climate," *Washington Post,* May 4, 2015.

8. See Chris P. Nielsen and Mun S. Ho, "Cleaning the Air in China," *New York Times,* Oct. 27, 2013. The authors, of the China Project at Harvard University's School of Engineering, are also the editors of *Clearer Skies over China: Reconciling Air Quality, Climate, and Economic Goals* (MIT Press, 2013).

9. I did that in spring 1999, as lead editor of a special report about coal combustion in China and India: William Sweet and Elizabeth A. Bretz, "Toward Carbon-Free Energy," *IEEE Spectrum,* Nov. 1, 1999, http://spectrum.ieee.org/energy/environment/toward-carbonfree-energy. Working with a China specialist, Marlowe Hood of AFP (Agence France Presse), I interviewed people at the Chinese Academy of Sciences, the State Power Corporation of China, the Electric Power Research Institute, the State Economic and Trade Commission, the Center for Renewable Energy Development, the Chinese Rural Energy Industries Association, the State Environmental Protection Commission, the Chinese Research Academy of Environmental Sciences, the Ministry of Agriculture, the UN Development Programme's Beijing office, the Clean Coal Engineering and Research Centre of Coal Industry, the Institute of Environmental Economics, the Beijing Environmental and Development Institute, and the China Energy and Environmental Technology Center.

10. Robert Orr, interview with the author, June 26, 2014.

11. Joby Warrick, "In Secret Talks, U.S., Chinese Officials Found Common Ground on Climate," *Washington Post,* Nov. 13, 2014. The lunch, which included US climate negotiator Stern and White House counselor Podesta, was with Yang Jiechi, a former foreign minister now serving as state councilor. Jeff Goodell, "The Secret Deal to Save the Planet," *Rolling Stone,* Dec. 9, 2014; John Podesta, interview with Al Hunt, Bloomberg TV, Nov. 13, 2014.

12. John Podesta, closing keynote address, National Summit on the Smart Grid and Climate Change, Washington, DC, Dec. 3, 2014.

13. See PBL Netherlands Environmental Assessment Agency and Joint Research Centre (of the European Commission), *Trends in Global CO₂ Emissions: 2014, Report,* 2014, table 2.2; David Fridley, Nina Zheng, and Yining Qin, *Inventory of China's Energy-Related CO₂ Emissions in 2008,* Lawrence Berkeley National Laboratory, LBNL-4600E, March 2011, table 1; Nan Zhou et al., *China's Energy and Carbon Emissions Outlook to 2050,* Lawrence Berkeley National Laboratory, LBNL-4472E, April 2011, figs. ES-1, ES-2; Xilian Zhang et al., *Carbon Emissions in China: How Far Can New Efforts Bend the Curve?,* MIT Joint Program on the Science and Policy of Climate Change, Report No. 267, October 2014, fig. 2. A significant source of the variation in estimates is whether they refer to all greenhouse gas emissions or just carbon emissions from fossil fuel use and cement production. One of the most recent estimates, in a Harvard study, puts 2012 emissions from fossil fuels and cement at 8.5 billion tons: Zhu Liu, *China's Carbon Emissions Report 2015,* Belfer Center for Science and International Affairs, Harvard Kennedy School, May 2015.

14. Using the Dutch numbers, for example, I postulated that China's emissions would grow 20% from 2015 to 2020, 10% from 2020 to 2025, and 5% from 2025 to 2030 (having grown 64% from 2000 to 2025 and 49% from 2002 to 2010): this means that its emissions would be 15.8 billion tons in 2030, compared to 3.65 billion tons (the Dutch figure) in 2000.

15. This is perhaps not surprising, considering how tough it has been to develop cost-effective substitutes for gasoline in vehicular transport. Even if electric cars were to largely replace those powered by internal combustion engines, carbon emissions would be significantly reduced only if most electricity were produced from zero-carbon sources.

16. The text of China's INDC is available at http://www4.unfccc.int/submissions/INDC/Published%20Documents/China/1/China's%20INDC%20-%20on%2030%20June%202015.pdf.

17. Orville Schell, "How China and U.S. Became Unlikely Partners on Climate," *Yale Environment 360,* Oct. 6, 2015, http://e360.yale.edu/feature/how_china_and_us_became_unlikely_partners_on_climate/2917/.

18. Narendra Modi, *Convenient Action: Gujarat's Response to Challenges of Climate Change* (Macmillan India, 2011). Hard copies of the book can be hard to come by, but it is readily located online and downloaded as a pdf.

19. See John H. Cushman Jr., "Obama Puts Climate Change High on the Agenda with Indian PM Modi," *InsideClimate News,* Oct. 2, 2014, http://insideclimatenews.org/carbon-copy/20141002/obama-puts-climate-change-high-agenda-indian-pm-modi; Niharika Mandhana, "U.S.-China Climate

Deal Puts India in Spotlight," *Wall Street Journal,* Nov. 17, 2014; and Peter Baker and Ellen Barry, "Obama Ends Visit with Challenge to India on Climate Change," *New York Times,* Jan. 27, 2015.

20. See Jairam Ramesh, minister of environment and forests from 2009 to 2011, "India's Call at Cancun Conclave," *Hindu,* June 30, 2014; and Chandrashekhar Dasgupta, "Raising the Heat on Climate Change," *Business Standard,* July 7, 2014.

21. Shyam Saran, "Leading at Paris," *Business Standard,* July 8, 2014.

22. Samar Saran, senior fellow and vice president, Observer Research Foundation, at India Project roundtable discussion, "Climate Change Negotiations and India," Brookings Institution, May 13, 2015.

23. See Ramana, *Power of Promise.*

24. Quoted in David Rose, "Why India Is Captured by Carbon," *Guardian,* May 27, 2015. The article provides a fine analysis of India's overall energy situation.

25. See Gardiner Harris, "Holding Your Breath in India," *New York Times,* May 31, 2015, and "Delhi Wakes Up to Problem It Cannot Ignore," *New York Times,* Feb. 15, 2015. In the May article, Harris worried that he had perhaps endangered the life of his eight-year-old son by having him in Delhi. An indignant public health expert wrote a letter to the paper in response, saying that Harris indeed had done so and had no business doing so. See also Katie Valentine, "India's Air Pollution Is Cutting Three Years Off the Lives of Its Residents," *Climate Progress,* Feb. 23, 2015, http://thinkprogress. org/climate/2015/02/23/3625819/india-deadly-air-pollution/.

26. Rakesh Mohan, interview with the author, Apr. 24, 2015.

27. Gupta, *History of Global Climate Governance,* 146.

28. Panjabi, *Earth Summit at Rio,* 178–79.

29. "India has more in common with least developed countries than with the emerging rapidly industrializing economies, but through its own negotiation strategies and external perception tends to be increasingly identified with the latter rather than the former, a point that has aroused criticism from long-standing allies in the G-77 such as Bangladesh." Navroz K. Dubash, "The Politics of Climate Change in India: Narratives of Equity and Co-Benefits," Centre for Policy Research, Climate Initiative, Working Paper 2012/1 (November 2012), New Delhi. The Centre for Policy Research is a major source for serious work on Indian climate policy and diplomacy. It is the only research organization in the world that, to my knowledge, has a program specifically dedicated to the study of climate diplomacy.

30. I rely here on the World Bank's data for per capita GDP and per capita CO_2 emissions, which are easily found online.

31. The text of India's INDC is available at http://www4.unfccc.int/submissions/INDC/Published%20Documents/India/1/INDIA%20INDC%20TO%20UNFCCC.pdf.

32. Victor Mallet, "Climate Change to Slash South Asian GDP, Development Bank Warns," *Financial Times*, Aug. 19, 2014. Business as usual would result in 1.8% lower growth by 2050 and 8.8% less by the end of the century, the Asian Development Bank found.

33. The closest Modi came to a statement about the environment was a promise to clean up the Ganges, so that Hindus practicing ritual bathing could swim in it safely.

34. A cynic would say, and I would agree, that Russia tries generally to be seen as a good citizen in all global matters it considers secondary to its national interest so that it can proceed ruthlessly in matters it deems primary: that means everything pertaining to its control of gas and oil export pipelines, as well as extension of its control over southeastern European regions deemed part of its historic sphere of interest—Georgia, Crimea, and eastern Ukraine, notably.

35. Gemenne, *Géopolitique du climate*, 33, 110; GHG data–UN-FCCC, Detailed data by Party, http://unfccc.int/di/DetailedByParty/Event.do?event=go.

36. Henry Sanderson and Pilita Clark, "Deripaska Warns about Competitive Risks from Paris Deal," *Financial Times*, Oct. 18, 2015.

37. Conversation, Trade Union Climate Summit, Murphy Institute, CUNY, June 29, 2015.

38. See, e.g., Rhett A. Butler, "Despite High Deforestation, Indonesia Making Progress on Forests, Says Norwegian Official," *Mongabay*, Oct. 2, 2014, http://news.mongabay.com/2014/10/despite-high-deforestation-indonesia-making-progress-on-forests-says-norwegian-official/.

39. "Which Countries Are Doing the Most to Stop Dangerous Global Warming?" *Guardian*, October 2015, http://www.theguardian.com/environment/ng-interactive/2015/oct/16/which-countries-are-doing-the-most-to-stop-dangerous-global-warming.

40. Bruce Jones and Samir Saran, "An 'India Exception' and India-US partnership on Climate Change: A Unique Dilemma," Brookings Institution, January 12, 2015, http://www.brookings.edu/blogs/planetpolicy/posts/2015/01/12-india-us-partnership-on-climate-change-jones-saran.

6
Sentimental Attachments, Existential Threats

1. The Group of 77 (G-77) at the United Nations, http://www.g77.org/doc/index.html.

2. Panjabi, *Earth Summit at Rio,* 5.

3. Panjabi, *Earth Summit at Rio,* 178.

4. José Goldemberg, "The Road to Rio," in Mintzer and Leonard, *Negotiating Climate Change,* 176–77.

5. Gupta, *History of Global Climate Governance,* 102. Joanna Depledge's assessment is even harsher: "The G-77 . . . has unswervingly refused to even discuss possible creative, moderate ways forward." Depledge "Opposite of Learning," 9.

6. Gupta, *History of Global Climate Governance,* 178.

7. Barnett, "Worst of Friends," 3. See also Depledge, "Opposite of Learning," 13.

8. Depledge, "Striving for No," 9.

9. Depledge, "Opposite of Learning," 12.

10. Depledge, "Striving for No," 19.

11. Chandrashekar Dasgupta, "The Climate Change Negotiations," in Mintzer and Leonard, *Negotiating Climate Change,* 135–36.

12. Benedick, *Ozone Diplomacy,* 326.

13. At the UN Climate Summit in September 2014, for example, the chair of the Alliance of Small Island States (AOSIS), Marlene Moses of Nauru, read a programmatic statement based on a scientists' report that AOSIS had commissioned. See "Tyndall Directors Author NY Climate Summit Report for the Alliance of Small Island States," *Tyndall Centre for Climate Change Research,* http://www.tyndall.ac.uk/communication/news-archive/2014/tyndall-directors-author-ny-climate-summit-report-alliance-small-isl.

14. A person from the Solomon Islands described this kind of action during a Global South session at the Climate Convergence Conference held in New York City on Sept. 20, 2014, on the eve of the UN Climate Summit.

15. According to Dasgupta, "Climate Change Negotiations," 135.

16. Gupta, *History of Global Climate Governance,* 186; Michele M. Betsill, "Environmental NGOs and the Kyoto Protocol Negotiations, 1995–1997," in Betsill and Corell, *NGO Diplomacy,* 52–53.

17. Tim Dickinson, "Canada's Crude Awakening," *Rolling Stone,* Mar. 12, 2015.

18. Michael L. Bloomberg, "Keystone Solution Runs through Canada," *BloombergView,* Feb. 25, 2015.

19. Diane Toomey, "How British Columbia Gained by Putting a Price on Carbon," *Yale Environment 360*, Apr. 30, 2015, http://e360.yale.edu/feature/how_british_columbia_gained_by_putting_a_price_on_carbon/2870/. Introduced in 2008, the revenue-neutral tax on carbon started at $10 per tonne and then rose by $5 per tonne during each of the next five years. Revenues were used to reduce income taxes, especially for lower-income people, and to provide some support to the carbon-intense cement industry. Suzanne Goldenberg, "Canada Switches on World's First Carbon Capture Facility," *Guardian*, Oct. 1, 2014.

20. Anne Pélouas, "Canada's Provinces Try to Exert Climate Change Pressure on Harper Government," *Guardian Weekly*, Apr. 28, 2015.

21. Pilita Clark, "Australia Marked Down for Reversal of Climate Change Law," *Financial Times*, Feb. 27, 2014.

22. See Australian Government, Department of Foreign Affairs and Trade, "Ambition Review under the Kyoto Protocol Second Commitment Period and Update on Australia's Greenhouse Gas Emissions Projections," May 2014, http://dfat.gov.au/international-relations/themes/climate-change/submissions/Pages/ambition-review-under-the-kyoto-protocol-second-commitment-period-and-update-on-australia-s-greenhouse-gas-emissions-projec.aspx. The country's emissions projections are at http://www.environment.gov.au/node/35053.

Also see "Australia Trounced Kyoto Climate Target, New Report Reveals," *Conversation*, Apr. 17, 2014, http://theconversation.com/australia-trounced-kyoto-climate-target-new-report-reveals-25744.

23. Lenore Taylor, "Australia Kills Off Carbon Tax," *Guardian*, July 16, 2014; Jamie Smyth, "Australia Risks Isolation among G20 by Scrapping Carbon Tax," *Financial Times*, June 29, 2014; James West, "One of the World's Worst Climate Villains Could Soon Be Booted from Office," *Mother Jones*, Feb. 4, 2015.

24. Alden Meyer worked first for the League of Conservation Voters, then for the Union of Concerned Scientists, which he joined in 1989. Meyer, interview with the author, Mar. 10, 2015.

25. See Betsill, "Environmental NGOs," 43–66.

26. Meyer interview.

27. Nick Buxton, "Cancun Agreement Stripped Bare by Bolivia's Dissent," *TNI*, Dec. 16, 2010, https://www.tni.org/en/article/cancun-agreement-stripped-bare-bolivias-dissent.

28. Hadden, *Networks in Contention*, 31–39. Between 1995 and 2015, according to Hadden's fig. 2.2, the number of NGOs showing up for COP meetings increased from fewer than 200 to more than 1,400. See also Fisher, "COP-15 in Copenhagen," 15.

29. Klein, *This Changes Everything*, 203–8, 219–20.

30. Klein, *This Changes Everything*, 207–8.

31. William Sweet, "Another Look at 'Deadly Gambits,'" *Bulletin of the Atomic Scientists* 41, no. 5 (1985): 49; Sweet, "Europe's Peace Movement: Topic or Target?" *Columbia Journalism Review* 22, no. 3 (1983), 46; Sweet, "Christian Peace Movement," *CQ Researcher,* May 13, 1983, http://library.cqpress.com/cqresearcher/document.php?id=cqresrre1983051300; Sweet, review of *From Protest to Policy: Beyond the Freeze to Common Security,* by Pam Solo, *Bulletin of the Atomic Scientists* 6, no. 2 (1989): 43–44.

32. Friedrich Engels to Joseph Bloch, Sept. 21, 1890, in *The Marx-Engels Reader,* ed. Robert C. Tucker (W. W. Norton, 1978), 761.

33. Richard Kinley, interview with the author, Apr. 14, 2014.

34. Gaetano Leone, interview with the author, Apr. 15, 2014.

35. See Andrew C. Revkin, "A Top Task for the New Chair of the UN Climate Panel—A Communications Reboot," *New York Times,* Oct. 6, 2015; and Ashley Rodriguez, "UN Climate Reports Are Becoming More Optimistic and Much Harder to Read," *Quartz,* Oct. 15, 2015, http://qz.com/527115/un-climate-reports-are-becoming-more-optimistic-and-much-harder-to-read/.

36. Quoted in Nick Cumming-Bruce, "UN Rights Chief Says He'll Shine a Light on Countries Big and Small," *New York Times,* Jan. 31, 2015.

37. See, e.g., Alan Holdren and Andrea Gagliarducci, "Full Transcript of Pope's Interview In-Flight to Manila," *Catholic News Agency,* Jan. 15, 2015.

38. Pope Francis, *Laudato Sí,* para. 169, 124–25.

7

The Road to Rio

1. Benedick, *Ozone Diplomacy,* 3.

2. See *Wikipedia,* s.v. "Montreal Protocol," https://en.wikipedia.org/wiki/Montreal_Protocol.

3. See Hoffman, *Ozone Depletion and Climate Change,* 105–22.

4. Benedick, *Ozone Diplomacy,* 3; Hoffman, *Ozone Depletion and Climate Change,* 93.

5. Hoffman, *Ozone Depletion and Climate Change,* 118–19.

6. Hoffman, *Ozone Depletion and Climate Change,* 92.

7. Benedick, *Ozone Diplomacy,* 67.

8. Benedick, *Ozone Diplomacy,* 57, 65.

9. G7 Summit, Toronto, June 19–21, 1988, conference declaration, para. 33.

10. Chandrashekhar Dasgupta, "The Climate Change Negotiations," in Mintzer and Leonard, *Negotiating Climate Change*, 130–31.

11. Benedick, *Ozone Diplomacy*, 1.

12. Brian Gareau, "Lessons from the Montreal Protocol in Phasing Out Methyl Bromide," *Journal of Environmental Studies and Sciences* 5, no. 2 (2015): 163.

13. Matthew Hoffman has an interesting discussion of this point. Hoffman, *Ozone*, 24.

14. See Oreskes and Conway, *Merchants of Doubt*.

8
Rio and Kyoto

1. Kyle W. Danish, "The International Regime," in Gerrard, *Global Climate Change and U.S. Law*, 33.

2. The convention text can be found at http://unfccc.int/files/essential_background/background_publications_htmlpdf/application/pdf/conveng.pdf.

3. The Organisation of Economic Cooperation and Development, consisting at that time of the United States, the western European nations, and the advanced capitalist economies of East Asia and the South Pacific.

4. Benedick, *Ozone Diplomacy*, 323.

5. William Nitze, for example, argued that had it been more forthcoming from the outset on stabilization, it might have sought instead to get more countries involved in making cuts. Nitze, "A Failure of Presidential Leadership," in Mintzer and Leonard, *Negotiating Climate Change*, 198.

6. As Joyeeta Gupta put it: "The rapidity with which the convention was drawn up and adopted is remarkable given the scientific complexity, the high economic stakes, the perceived abstract nature of the future impacts, and the differing interests of the global community." Gupta, *History of Global Climate Governance*, 87.

7. In the last ten days of INC negotiations, the North largely resolved its internal differences and agreement was reached on a draft treaty that would confine emissions commitments to the industrial countries, enunciated the principle of common but differentiated responsibility, recognized that developing country contributions would depend on outside assistance,

and did *not* call for review of LDC programs. Dasgupta, "The Climate Change Negotiations," in Mintzer and Leonard, *Negotiating Climate Change*, 129–48.

8. Gupta, *History of Global Climate Governance*, 85.

9. There are two authoritative accounts: Mary Elise Sarotte, *1989: The Struggle to Create Post–Cold War Europe* (Princeton University Press, 2009), and Philip Zelikow and Condoleezza Rice, *Germany Unified and Europe Transformed: A Study in Statecraft* (Harvard University Press, 1995).

10. Michele M. Betsill, "Environmental NGOs and the Kyoto Protocol Negotiations, 1995–97," in Betsill and Corell, *NGO Diplomacy*, 50–51.

11. Betsill, "Environmental NGOs," 53.

12. The targeted cuts in this first Kyoto compliance period referred to each party's average annual emissions in the five years from 2008 to 2012.

13. Those fears proved to be well founded. Shortly after George W. Bush took office, his national security adviser declared Kyoto to be "dead" as far as the United States was concerned. Six years later Japan declared it would not take part in the second Kyoto commitment period, in which industrial country cuts were to be negotiated for 2012–20. Following the derailment of the Kyoto process at Copenhagen in December 2009, first Canada and then New Zealand withdrew from the protocol. Thus, of the original US negotiating allies, only Australia—beset by chronic devastating wildfires—remained aboard Kyoto.

14. Entry into force required at least fifty-five countries accounting for at least 55 percent of global emissions to ratify the protocol. In effect, either the United States or Russia, or both, would have to ratify the treaty for it to enter into force. It was a close call. By 2005, because of the US failure to ratify, serious doubts had developed in Russia as to whether much money was to be made selling emissions credits. See Anna Korppoo, Jacqueline Karas, and Michael Grubb, eds., *Russia and the Kyoto Protocol: Opportunities and Challenges* (Chatham House, 2006), 70.

15. William Nordhaus, "After Kyoto: Alternative Mechanisms to Control Global Warming," *AEA Papers and Proceedings* 96, no. 2 (2006): 32.

16. Richard N. Cooper, "The Kyoto Protocol: A Flawed Concept," *Environmental Law Reporter* 31, no. 12 (2001): 145–72.

17. Richard N. Cooper, "Alternatives to Kyoto: The Case for a Carbon Tax," Weatherhead Center for International Affairs, Harvard University, 2006, http://wcfia.harvard.edu/publications/alternatives-kyoto-case-carbon-tax.

18. Yet Nordhaus also expressed a somewhat rueful attitude about the apparent demise of Kyoto, in *Climate Casino*, 248: "It is painful to conclude that the important and well-meaning approach—in which so many invested so much time and hope—has failed."

19. Gerrard, *Global Climate Change and U.S. Law,* 53.

20. Dessler and Parson, *Science and Politics of Global Climate Change,* 164.

21. Fehl, *Living with a Reluctant Hegemon,* 32.

22. IPCC, *Fifth Assessment Report (AR5),* chap. 13, 60. That number accounts for changes in land use and forestry. Without that the decrease was 16.6%. See also Aldy, "Policy Surveillance in the G-20 Fossil Fuel Subsidies Agreement," 7, who says that industrial country emissions (including US emissions) were down 10–15% from 1990 to 2008–12 (greater than required), so that on its face "the Kyoto Protocol was a success."

23. European Environment Agency, "Tracking Progress towards Kyoto and 2020 Targets in Europe," EEA Report No. 7/2010, p. 6.

24. Corina Haita, "The State of Compliance in the Kyoto Protocol," ICCG, Reflection No. 12/2012, p. 3.

25. European Environment Agency, "Why Did GHG Emissions Decrease in the EU between 1990 and 2012?" June 2014.

26. See William Sweet, "Restoring Coal's Sheen: Swedish Energy Company Takes a Novel Approach to Carbon Capture," *IEEE Spectrum,* Jan. 1, 2008, http://spectrum.ieee.org/green-tech/clean-coal/winner-restoring-coals-sheen.

27. Although the overall results are not spectacular, they are not terrible either. To say "very few" countries were in compliance does not accurately reflect the facts.

28. Fehl, *Living with a Reluctant Hegemon,* 133.

29. Connecting the Framework Convention with the Nuclear Non-Proliferation Treaty may seem "out of left field," as the saying goes. But Naomi Klein makes the connection too, in *This Changes Everything,* 23, though she does not develop it.

30. Albert Wohlstetter, "Spreading the Bomb without Quite Breaking the Rules," *Foreign Policy* 25 (Winter 1976–77): 88–96, 145–79. See also William Sweet, *The Nuclear Age: Energy, Proliferation, and the Arms Race* (Congressional Quarterly, 1984), 150.

31. See the Nuclear Threat Initiative's country profile for South Africa: http://www.nti.org/country-profiles/south-africa/nuclear/.

32. See Jay C. Davis and David A. Kay, "Iraq's Secret Nuclear Weapons Program," *Physics Today,* July 1992, 21–27. Inexplicably, the administration of George W. Bush refused to recognize that the United Nations had done its job successfully and insisted on invading Iraq a second time, with catastrophic results.

33. See David Albright, *Peddling Peril: How the Secret Nuclear Trade Arms America's Enemies* (Free Press, 2010).

34. See William Sweet, "Iran's Nuclear Program Reaches Critical Juncture," *IEEE Spectrum,* June 2004, 12–15.

35. On Iraq: Unless, in this counterhistory, Israel acted unilaterally to knock out Saddam's nuclear infrastructure. Israel was in fact the first to take military action against Iraq, destroying a reactor that Saddam Hussein was building with French assistance in 1979.

36. Tellingly, when a trio of top economists turned their attention to climate change, in the *Journal of Economic Literature* 51, no. 3 [2014]: 838–882, all three referred to "climate catastrophe" prominently in their abstracts. The three articles are: Nicholas Stern, "The Structure of Economic Modeling of the Potential Impacts of Climate Change: Grafting Gross Underestimation of Risk onto Already Narrow Science Models"; Robert S. Pindyck, "Climate Change Policy: What Do the Models Tell Us?"; and Martin L. Weitzman, "Tail-Hedge Discounting and the Social Cost of Carbon." See also Oreskes and Conway, "Collapse of Western Civilization."

37. See www.trillionthtonne.org; see also Fred Pearce, "The Trillion-Ton Cap: Allocating the World's Carbon Emissions," *Yale Environment 360,* Oct. 24, 2013, http://e360.yale.edu/feature/the_trillion-ton_cap_allocating_the_worlds_carbon_emissions/2703/.

38. UNEP, *The Emissions Gap Report 2013: A UNEP Synthesis Report,* http://www.unep.org/publications/ebooks/emissionsgapreport2013/. Oliver Milman, "Carbon Emissions Must Be Cut 'Significantly' by 2020, Says UN Report," *Guardian,* Nov. 5, 2013.

39. "Shift to a New Climate Likely by Mid-Century—Study," Reuters, Oct. 9, 2013.

9

Copenhagen

1. In a short videotaped speech that President-Elect Obama delivered to the Bipartisan Governors Climate Summit on November 18, 2008, he said that from now on, any US governor seeking to reduce emissions and any company seeking to develop green technology would have an ally in the White House.

2. The full text of the letter is at http://www.350.org/breaking-powerful-appeal-desmond-tutu.

3. Evidently the draft treaty had been prepared by a number of major governments and Denmark's incumbent, rather conservative government, which was generally much warier of taking strong action on climate than its environment minister Connie Hedegaard, who initially was chairing the conference. For the contents of the draft and its political impact, see John Vidal, "Copenhagen Climate Summit in Disarray after 'Danish Text' Leak," *Guardian,* Dec. 8, 2009.

4. I was among them, having obtained accreditation to attend as a member of an NGO.

5. See Fisher, "COP-15 in Copenhagen"; and Hadden, *Networks in Contention.*

6. *Earth Negotiations Bulletin* 12, no. 459 (Dec. 22, 2009), 2.

7. Bill Sweet, "Constructive Ideas for Copenhagen Climate Conference," *IEEE Spectrum,* Sept. 24, 2009, http://spectrum.ieee.org/energywise/energy/renewables/constructive-ideas-for-copenhagen-conference.

8. This in fact is what the Obama administration ended up doing.

9. The full text of the three-page document is at: http://unfccc.int/resource/docs/2009/cop15/eng/11a01.pdf.

10. For Stern's various statements, see e.g., Darren Samuelsohn, "Senate's Climate Blueprint Raises Expectations for Obama on World Stage," *New York Times,* Dec. 11, 2009; Andrew C. Revkin and Tom Zeller Jr., "U.S. Negotiator Dismisses Reparations for Climate," *New York Times,* Dec. 10, 2009; Elisabeth Rosenthal and Neil MacFarquhar, "Industrialized Nations Unveil Plans to Rein in Emissions," *New York Times,* Nov. 20, 2009; and the video of Stern's press conference on Dec. 9, 2009, at https://www.youtube.com/watch?v=FQq3MQSfTZg.

11. Five years later, an organization had been established in Inchon, South Korea, to manage the fund and disburse money, but for all practical purposes no national allocations had been made to the fund, and it was not at all clear how or when such allocations might be made. "Most see the faint mention of sources as unsurprising," commented analyst Richard K. Lattanzio drily, in his "International Climate Change Financing: The Green Climate Fund," Congressional Research Service Report R41889, Apr. 16, 2013, 8.

12. The text of the Bali Action plan can be viewed at http://unfccc.int/resource/docs/2007/cop13/eng/06a01.pdf.

13. Lula's ten-minute talk can be viewed with subtitles on YouTube at: https://www.youtube.com/watch?v=LQzVjDp5WA8.

14. Shapiro, *Carbon Shock,* 176.

15. World Bank Data, "CO Emissions (metric tons per capita)," http://data.worldbank.org/indicator/EN.ATM.CO2E.PC. France's 2010 per capita GHG emissions were 5.6 metric tons.

16. World Bank Data, "CO Emissions (metric tons per capita)." Generally, only the top oil-producing countries of the Middle East and Central Asia had even higher per capita emissions.

17. The 27 current members of the European Union, for example, promised 20–30% cuts from 1990 levels by the year 2020; Japan proposed similar cuts of 25–32%, and South Korea 4%. Russia pledged cuts of 4–15% by 2020, on top of a 34% reduction already registered. (Having seen its economy implode in the 1990s and having then adopted some improved technology with its economic recovery, Russia's gains were occurring almost matter-of-course. Much the same was true of other former Soviet Bloc and Eastern European countries, which had received special treatment because of their unique situation in the Kyoto Protocol.) Brazil said that its 2020 emissions would be 36–39% lower than otherwise projected, and China aimed for a 40–45% improvement in carbon. India and South Africa stood out among the industrializing countries in making no commitments whatsoever. Australia, whose emissions were 30% above 1990 levels, promised only a 5% cut. Canada, with emissions 26% above 1990, proposed a 3% cut. In this particular context, the United States, proposing a 17% or 17.5% reduction, looked a good deal better. But its proposed cut referred to 2005, not 1990, as the baseline and still would leave 2020 emissions far above what Kyoto had called for. See "Climate Pledges So Far Nowhere Near Enough," *New Scientist*, Oct. 17, 2009, 14–15; Tim Johnson, "Climate Change Talks Fail to Break Impasse," *Financial Times*, Oct. 3, 2009, 13; and Elisabeth Rosenthal and Neil MacFarquhar, "Nations Unveil Plans to Rein in Emissions," *New York Times*, Nov. 20, 2009.

18. The transcript of the speech Obama delivered at Georgetown University on June 25, 2013, is at http://www.whitehouse.gov/the-press-office/2013/06/25/remarks-president-climate-change.

19. *Earth Negotiations Bulletin* 12, no. 459 (Dec. 22, 2009): 27.

20. Suzanne Goldenberg, "Obama Must Pass Climate Laws ahead of Copenhagen, Danish Minister Warns," *Guardian*, Mar. 3, 2009.

21. Dessler and Parson, *Science and Politics of Global Climate Change*, 29.

22. Alf Wills, interview with the author, Dec. 1, 2015.

23. Mark Lynas, "How Do I Know China Wrecked the Copenhagen Deal? I Was in the Room," *Guardian*, Dec. 22, 2009.

24. Mark Lynas, blog post, "Why It's Wrong to Preach 'Climate Justice,'" *New Statesman*, Jan. 14, 2010, http://www.newstatesman.com/blogs/the-staggers/2010/01/lynas-climate-carbon.

25. http://www.theguardian.com/environment/blog/2009/dec/18/copenhagen-climate-change-summit-liveblog/ (link no longer active).

26. David Corn and Kate Sheppard, "Obama's Copenhagen Deal," *Mother Jones,* Dec. 18, 2009, http://www.motherjones.com/environment/2009/12/obamas-copenhagen-deal.

27. Carlarne, *Climate Change and Policy,* 352–53.

28. Klein, *This Changes Everything,* 18–19.

29. Saran, "Irresistible Forces and Immovable Objects: A Debate on Contemporary Climate Politics," *Climate Policy* 10, no. 6 (2010): 680.

30. "Dr. Surya Sethi on the Barcelona Climate Discussions: Stalemate," *NewsClickin,* https://www.youtube.com/watch?v=h-W0YUQLOfg.

31. Surya Sethi, interview with the author, July 8, 2014.

10
The Road to Paris

1. Richard Kinley, interview with the author, Apr. 14, 2014.

2. Elements of the Durban Platform, which the Framework Convention on Climate Change's secretariat describes as a diplomatic breakthrough, are enumerated and explained on the secretariat's website: http://unfccc.int/meetings/durban_nov_2011/meeting/6245/php/view/decisions.php.

3. According to Alden Meyer of the Union of Concerned Scientists, who had attended all the Conferences of the Parties but one.

4. See, e.g., Marlowe Hood, "France Signals 'Breakthrough' in Climate Talks," Agence France Presse, July 21, 2015; and Pilita Clark, "Climate Talks Advancing Faster towards December Deal," *Financial Times,* July 19, 2015.

5. Bodansky, "United Nations Framework Convention on Climate Change."

6. Their aggregation of Paris pledges represents an extrapolation from current energy and climate policies worldwide. They also have a third scenario, a "more optimistic extended Copenhagen scenario," which I am disregarding here for simplicity's sake.

7. Jacoby and Chen, "Expectations for a New Climate Agreement," 18, table 1.2.

8. Jacoby and Chen, "Expectations for a New Climate Agreement," 21.

9. Jacoby and Chen, "Expectations for a New Climate Agreement," 15–16.

10. Jacoby and Chen, "Expectations for a New Climate Agreement," figs. 4–6.

11. Bodansky and Diringer, "Building Flexibility and Ambition into a 2015 Climate Agreement," 1.

12. Bodansky and Diringer, "Building Flexibility and Ambition into a 2015 Climate Agreement," 5, 11.

13. Bodansky and Diringer, "Building Flexibility and Ambition into a 2015 Climate Agreement," 13.

14. At a forum held under the auspices of the Yale School of Forestry and Environmental Sciences on April 10, 2015, to discuss the implications of the papal climate encyclical, Yale Law School professor Douglas Kysar said: "The global climate policy process has been held hostage by a country, this country, which has found its own political process held hostage by economic interests that are capable of investing not only in traditional capital but also in the capture of laws and institutions that are intended to regulate capital." A transcript of the event is available at http://fore.yale.edu/files/Papal_Panel_Transcript.pdf.

15. Bodansky, "Legal Options for U.S. Acceptance of a New Climate Change Agreement," 1. See also his "Key Legal Issues in the 2015 Climate Negotiations."

16. Henry D. Jacoby, interview with the author, Sept. 8, 2014. Jacoby noted that in addition to their general promises to cut emissions 20% by 2020 and 40% by 2030, the Europeans have set strong targets for ETS, amounting to 1.7–2.12% emissions cuts per year.

17. The language of the commitment itself was rather slippery. Paragraph 8 of the Copenhagen Accord says that the US$100 billion "will come from a wide variety of sources, public and private, bilateral and multilateral, including alternative sources of finance."

18. Daniel Bodansky, e-mail to author, Jan. 15, 2015.

19. Selwyn C. Hart, lecture, New School, New York, May 11, 2015.

Selected Bibliography

Aldy, Joseph E. "Policy Surveillance in the G-20 Fossil Fuel Subsidies Agreement: Lessons for Climate Policy." HKS Faculty Research Working Paper Series RWP15-029, June 2015.

Aldy, Joseph E., and Robert N. Stavins, eds. *Post-Kyoto International Climate Policy: Implementing Architectures for Agreement.* Cambridge University Press, 2010.

Allen, Myles R., and Thomas F. Stocker. "Impact of Delay in Reducing Carbon Dioxide Emissions." *Nature Climate Change* (2013), doi:10.1038/nclimate2077.

Aykut, Stefan, and Amy Dahan. *Gouverner le climat: 20 ans de négotiations internationales.* Presses de Sciences Po, 2014.

Balibar, Sébastiaen. *Climat: Y voir clair pour agir.* Le Pommier, 2015.

Barnes, Peter, and Marc Breslow. *Pie in the Sky? The Battle for Atmospheric Scarcity Rents.* Political Economy Research Institute, Working Paper Series No. 13, University of Massachusetts, Amherst, 2001.

Barnett, Jon. "The Worst of Friends: OPEC and G-77 in the Climate Regime." *Global Environmental Politics* 8, no. 4 (2008): 1–8.

Bauer, Steffen. "It's about Development, Stupid! International Climate Policy in a Changing World." *Global Environmental Politics* 12, no. 2 (2012): 110–15.

Benedick, Richard E. "Avoiding Gridlock on Climate Change." *Issues in Science and Technology* 23, no. 2 (2007): 37–40.

———. *Ozone Diplomacy: New Directions in Safeguarding the Planet.* Harvard University Press, 1991, enlarged ed. 1998.

Betsill, Michele M., and Elisabeth Corell, eds. *NGO Diplomacy: The Influence of Nongovernmental Organizations in International Environmental Negotiations.* MIT Press, 2008.

Better Growth, Better Climate. The New Climate Economy, The Global Commission on the Economy and Climate, September 2014.

Beyond 2015: An Innovation-Based Framework for Global Climate Policy. Center for Clean Energy Innovation, Washington, DC, 2005.

Bodansky, Daniel. "Bonn Voyage: Kyoto's Uncertain Revival." *National Interest* 65 (Fall 2001): 45–56.

———. "Key Legal Issues in the 2015 Climate Negotiations." Center for Climate and Energy Solutions, June 2015.

———. "Legal Options for U.S. Acceptance of a New Climate Change Agreement." Center for Climate and Energy Solutions, May 2015.

———. "The United Nations Framework Convention on Climate Change: A Commentary." *Yale Journal of International Law* 18, no. 2 (1993): 451–558.

Bodansky, Daniel, and Elliot Diringer. "Building Flexibility and Ambition into a 2015 Climate Agreement." Center for Climate and Energy Solutions, June 2014.

Bosetti, Valentina, and Jeffrey Frankel. *A Pre-Lima Scorecard for Evaluating Which Countries Are Doing Their Fair Share in Pledged Carbon Cuts.* Harvard Project on Climate Agreements, November 2014.

Canfin, Pascal, and Peter Staime. *Le Climat: 30 questions pour comprendre la Conférence de Paris.* Les Petits Matins, 2015.

Carlarne, Cinnamon P. *Climate Change Law and Policy: EU and US Perspectives.* Oxford University Press, 2010.

Carlarne, Cinnamon P., Kevin R. Gray, and Richard Tarasofsky. *The Oxford Handbook of Climate Change Law.* Oxford University Press, 2016.

Chakravarty, Shoibal, et al. "Sharing Global CO_2 Emission Reductions among One Billion High Emitters." *PNAS* 106, no. 29 (2009): 11884–88.

Davis, Steven J., and Robert H. Socolow. "Commitment Accounting of CO_2 Emissions," *Environment Research Letters* 9, no. 8 (2014), doi:10.1088/1748-9326/9/8/084018.

Depledge, Joanna. "The Opposite of Learning: Ossification in the Climate Change Regime." *Global Environmental Politics* 6, no. 1 (February 2006): 1–22.

———. "Striving for No: Saudi Arabia in the Climate Change Regime." *Global Environmental Politics* 8, no. 4 (2008): 9–35.

Dessler, Andrew E., and Edward A. Parson. *The Science and Politics of Global Climate Change: A Guide to the Debate,* 2nd ed. Cambridge University Press, 2011.

Dimitrov, Radoslav S. "Inside Copenhagen: The State of Climate Governance." *Global Environmental Politics* 10, no. 2 (2010): 18–24.

Eckersley, Robyn. "Moving Forward in the Climate Negotiations: Multilateralism or Minilateralism?" *Global Environmental Politics* 12, no. 2 (2012): 24–42.

The Economics of Climate Change Mitigation: Policies and Options for Global Action beyond 2012. OECD, September 2009.

Elliott, Lorraine. "Climate Diplomacy." In *The Oxford Handbook of Modern Diplomacy,* ed. Andrew F. Cooper, Jorge Heine, and Ramesh Thakur, 840–56. Oxford University Press, 2013.

Energy Technology Perspectives: Harnessing Electricity's Potential. International Energy Agency, 2014 and 2015.

Fagan, Brian. *Elixir: A History of Water and Humankind.* Bloomsbury, 2011.

Fehl, Caroline. *Living with a Reluctant Hegemon: Explaining European Responses to U.S. Unilateralism.* Oxford University Press, 2012.

Fisher, Dana R. "COP-15 in Copenhagen: How the Merging of Movements Left Civil Society Out in the Cold." *Global Environmental Politics* 10, no. 2 (2010): 11–17.

Flannery, Tim. *Atmosphere of Hope: Searching for Solutions to the Climate Crisis.* Atlantic Monthly Press, 2015.

————. *The Weather Makers: The History and Future Impact of Climate Change.* Atlantic Monthly Press, 2005

Gardiner, Stephen M., Simon Caney, Dale Jamieson, and Henry Shue, eds. *Climate Ethics: Essential Readings.* Oxford University Press, 2010.

Gemenne, François. *Géopolitique du climate: Négociations, stratégies, impacts.* Armand Colin, 2015.

Gerrard, Michael B., ed. *Global Climate Change and U.S. Law.* American Bar Association, Section of Environment, Energy and Resources, 2007.

Giddens, Anthony. *The Politics of Climate Change,* 2nd ed. Polity Press, 2011.

Goodell, Jeff. *Big Coal: The Dirty Secret behind America's Energy Future.* Houghton Mifflin, 2006.

Gore, Al. *An Inconvenient Truth.* Rodale, 2006.

Green, Jessica F. *Rethinking Private Authority: Agents and Entrepreneurs in Global Environmental Governance.* Princeton University Press, 2015.

Grubb, Michael, et al., eds. *The "Earth Summit" Agreements: A Guide and Assessment.* Earthscan, 1993.

Gupta, Joyeeta. *The History of Global Climate Governance.* Cambridge University Press, 2014.

Gupta, Joyeeta, and Nicolien van der Grijp, eds. *Mainstreaming Climate Change in Development Cooperation.* Cambridge University Press, 2010.

Gupta, Joyeeta, and Michael Grubb. *Climate Change and European Leadership: A Sustainable Role for Europe?* Kluwer Academic, 2000.

Hadden, Jennifer. *Networks in Contention: The Divisive Politics of Climate Change.* Cambridge University Press, 2015.

Held, David, Angus Fane-Hervey, and Marika Theros, eds. *The Governance of Climate Change: Science, Economics, Politics and Ethics.* Polity, 2011.

Hoffman, Matthew J. *Climate Governance at the Crossroads: Experimenting with a Global Response after Kyoto.* Oxford University Press, 2011.

———. *Ozone Depletion and Climate Change: Constructing a Global Response.* SUNY Press, 2005.

Intergovernmental Panel on Climate Change (IPCC). *Fifth Assessment Report (AR5).* IPCC, Working Groups 1, 2, and 3, 2013 and 2014.

Jacoby, Henry D., and Y.-H. Henry Chen. "Expectations for a New Climate Agreement." MIT Joint Program on the Science and Policy of Global Change, Report No. 264, August 2014.

Keohane, Robert O., and David G. Victor. "The Regime Complex for Climate Change." Harvard Kennedy School, Harvard Project on International Climate Agreements, Discussion Paper 10-33, January 2010.

Klare, Michael T. *The Race for What's Left: The Global Scramble for the World's Last Resources.* Metropolitan Books, 2012.

———. *Rising Powers, Shrinking Planet.* Henry Holt, 2008.

Kleber, Claus, and Cleo Paskal. *Spielball Erde: Machtkämpfe im Klimawandel.* C. Bertelsmann, 2012.

Klein, Naomi. *This Changes Everything: Capitalism vs. the Climate.* Simon and Schuster, 2014.

———. *Field Notes from a Catastrophe: Man, Nature and Climate Change.* Bloomsbury, 2007.

Klein, Naomi, with reply by Elizabeth Kolbert. "Can Climate Change Cure Capitalism? An Exchange." *New York Review of Books,* Jan. 8, 2015.

Lovins, Amory, and Marvin Odum. *Reinventing Fire: Bold Business Solutions for the New Business Era.* Chelsea Green, 2011.

Lovins, L. Hunter, and Boyd Cohen. *Climate Capitalism: Capitalism in the Age of Climate Change.* Hill and Wang, 2011.

Lutes, Mark. *Emerging Leaders: How the Developing World Is Starting a New Era of Climate Change Leadership.* World Wildlife Fund, November 2009.

Meckling, Jonas. "The Globalization of Carbon Trading: Transnational Business Coalitions in Climate Politics." *Global Environmental Politics* 11, no. 2 (2011): 26–50.

Merrill, Laura, Andrea M. Bassi, Richard Bridle, and Lasse T. Christensen. *Tackling Fossil Fuel Subsidies and Climate Change: Levelling the Energy Playing Field.* Nordic Council of Ministers (Norden), 2015.

Mintzer, Irving M., and J. Amber Leonard, eds. *Negotiating Climate Change: The Inside Story of the Rio Convention.* Cambridge University Press, 1994.

Mistral, Jacques, ed. *Le climat va-t-il changer le capitalisme? La grande mutation du XXIe siècle.* Groupe Eyrolles, 2015.

Mitigation Goal Standard: An Accounting and Reporting Standard for National and Subnational Greenhouse Gas Reduction Goals. Greenhouse Gas Protocol, World Resources Institute, November 2014.

Moosa, Valli, and Harald Dovland. *Vision for Paris: Building an Effective Climate Agreement.* Center for Climate and Energy Solutions, July 2015.

Newell, Peter, and Matthew Paterson. *Climate Capitalism: Global Warming and the Transformation of the Global Economy.* Cambridge University Press, 2010.

Nordhaus, William. *The Climate Casino: Risk, Uncertainty, and Economics for a Warming World.* Yale University Press, 2013.

———. "Climate Clubs: Overcoming Free-Riding in International Climate Policy." *American Economic Review* 104, no. 2 (2015): 1339–70.

———. "A New Solution: The Climate Club." *New York Review of Books,* June 14, 2015.

Oglesby, Donna Marie. *Spectacle in Copenhagen: Public Diplomacy on Parade.* USC Center on Public Diplomacy, Figueroa Press, December 2010.

Oreskes, Naomi, and Erik M. Conway. "The Collapse of Western Civilization: A View from the Future." *Daedalus* 142, no. 1 (2013): 40–58.

———. *The Collapse of Western Civilization: A View from the Future.* Columbia University Press, 2014.

———. *Merchants of Doubt: How a Handful of Scientists Obscured the Truth on Issues from Tobacco Smoke to Global Warming.* Bloomsbury Press, 2010.

Orr, Shannon K. "Reimagining Global Climate Change: Alternatives to the UN Treaty Process." *Global Environmental Politics* 11, no. 4 (2011): 134–38.

Page, Edward R. *Climate Change, Justice and Future Generations.* Edward Elgar, 2006.

Panjabi, Ranee K. L. *The Earth Summit at Rio: Politics, Economics, and the Environment.* Northeastern University Press, 1997.

Pope Francis. *Laudato Sí: On Care for Our Common Home.* Papal encyclical, Vatican, 2015.

Posner, Eric A., and David Weisbach. *Climate Change Justice.* Princeton University Press, 2010.

Prins, Gwyn, and Steve Rayner. *The Wrong Trousers: Radically Rethinking Climate Policy.* Joint discussion paper, James Martin Institute for

Science and Civilization, University of Oxford, and London School of Economics, 2007.

Purvis, Nigel, and Andrew Stevenson. *Rethinking Climate Diplomacy: New Ideas for Transatlantic Cooperation Post-Copenhagen.* Brussels Forum Paper Series, German Marshall Fund of the United States, March 2010.

Rahmstorf, Stefan, and Hans Joachim Schellnhuber. *Der Klimawandel: Diagnose, Prognose, Therapie.* C. H. Beck, 2006.

Rajamani, Lavanya. "Differentiation in a 2015 Climate Agreement." Center for Climate and Energy Solutions, June 2015.

Ramana, M. V. *The Power of Promise: Examining Nuclear Energy in India.* Viking/Penguin Books, 2012.

Ramana, M. V., and Zia Mian. "One Size Doesn't Fit All: Social Priorities and Technical Conflicts for Small Modular Reactors." *Energy Research and Social Science* 2 (June 2014): 115–24.

Ramana, M. V., and Eri Salkawa. "Choosing a Standard Reactor: International Competition and Domestic Politics in Chinese Nuclear Policy." *Energy* 36, no. 12 (2011): 6779–89.

Rawls, John. *A Theory of Justice.* Rev. ed. Belknap Press of Harvard University Press, 1999.

Roberts, J. Timmons, et al. "Who Ratifies Environmental Treaties and Why?" *Global Environmental Politics* 4, no. 3 (2004): 22–63.

Rockstrom, Johan, and Mattias Klum. *Big World, Small Planet: Abundance within Planetary Boundaries.* Yale University Press, 2014.

Sachs, Jeffrey D. *The Age of Sustainable Development.* Columbia University Press, 2015.

Sachs, Jeffrey, et al. *Pathways to Deep Decarbonization.* Sustainable Development Solutions Network and Institute for Sustainable Development and International Relations, 2014.

Sandor, Richard L. *Good Derivatives: A Story of Financial and Environmental Innovation.* Wiley, 2012.

Schäfer, Andreas, et al. *Transportation in a Climate-Constrained World.* MIT Press, 2009.

Schellnhuber, Hans Joachim. *Selbstverbrennung: Die fatale Dreiecksbeziehung zwischen Klima, Mensch, und Kohlenstoff.* C. Bertelsmann, 2015.

Schneider, Stephen H. *Science as a Contact Sport: Inside the Battle to Save Earth's Climate.* National Geographic Society, 2009.

Shapiro, Mark. *Carbon Shock: A Tale of Risk and Calculus on the Front Lines of a Disrupted Global Economy; Discovering the New Normal of a Climate-Impacted World.* Chelsea Green, 2014.

Shelling, Thomas. "What Makes Greenhouse Sense?" *Foreign Affairs* 81, no. 3 (2002): 2–9.

Sperling, Daniel, and James S. Cannon, eds. *Driving Climate Change: Cutting Carbon from Transportation.* Elsevier, 2007.

Speth, James Gustave. *Red Sky at Morning: America and the Crisis of the Global Environment,* 2nd ed. Yale University Press, 2005.

Stavins, Robert N. "The Problem of the Commons: Still Unsettled after 100 Years." *American Economic Review* 101, no. 1 (2011): 81–108.

Stern, Nicholas. "The Economics of Climate Change." *American Economic Review* 98, no. 2 (2008): 1–37.

———. *The Global Deal: Climate Change and the Creation of a New Era of Progress and Prosperity.* Public Affairs, 2009.

Stern, Nicholas, et al. *Stern Review on the Economics of Climate Change.* London, 2006.

Sweet, William. *Kicking the Carbon Habit: The Case for Renewable and Nuclear Energy.* Columbia University Press, 2006.

UNEP. *The Emissions Gap Report 2014: A UNEP Synthesis Report.* United Nations Environment Programme, November 2014.

Victor, David G. *Climate Change: Debating America's Policy Options.* A Cornell Policy Initiative, sponsored by the Council on Foreign Relations, distributed by the Brookings Institution, 2004.

———. *Global Warming Gridlock: Creating More Effective Strategies for Protecting the Planet.* Cambridge University Press, 2015.

Wiener, Jonathan B. "Think Globally, Act Globally: The Limits of Local Climate Policies." *University of Pennsylvania Law Review* 155, no. 6 (2007): 1961–79.

Zedillo, Ernesto, ed. *Global Warming: Looking beyond Kyoto.* Brookings Institution Press, 2008.

Acknowledgments

I am grateful to the following individuals for sharing thoughts, insights, and information with me, whether in interviews, quick conversations, or e-mail exchanges. None is quoted or cited in the book without the person's express, advance permission.

Daniel Bodansky, professor of law, Sandra Day O'Connor College of Law, Arizona State University, and former US climate negotiator

Cinnamon Carlarne, professor of law, Moritz College of Law, Ohio State University

David Courard-Hauri, professor of environmental science and policy, Drake University

Paula J. Dobriansky, senior fellow, Belfer Center for Science and International Affairs, Harvard Kennedy School, and former undersecretary of state for democracy and global affairs; member of U.S. delegation, 2007 Bali climate conference

David Fridley, staff scientist, China Energy Group, Lawrence Berkeley National Laboratory

Peter Haas, professor of political science, University of Massachusetts, Amherst

Jeffrey Hayward, director, Climate Program, Rainforest Alliance

Henry D. Jacoby, professor emeritus, MIT Sloan School of Management, former director of the Harvard Environmental Systems Program, and former director of the MIT Center for Energy and Environmental Policy Research

Richard Kinley, Deputy Executive Secretary, Climate Change Secretariat (UNFCCC), Bonn

Reto Knutti, leader, climate physics group, ETH Zurich

Kerstin Krellenberg, Helmholtz-Zentrum für Umweltforschung, Leipzig

Gaetano Leone, former deputy secretary, IPCC, Geneva

Anthony Leiserowitz, director, Yale Project on Climate Change Communication, Yale School of Forestry and Environmental Studies

Edward Maibach, professor of communication and director, Center for Climate Change Communication, George Mason University

Alden Meyer, director of strategy and policy, Union of Concerned Scientists

Rakesh Mohan, member, Governing Board, International Monetary Fund

Alexander Nauels, doctoral student, Australian-German Climate and Energy College, Melbourne, and former technical unit member, IPCC Working Group I for fifth assessment report

Michael Oppenheimer, professor of geosciences and international affairs, Princeton University

Robert Orr, assistant secretary-general for policy, United Nations, New York

Nigel Purvis, founding president and CEO, Climate Advisers, and former senior-level US climate diplomat

M. V. Ramana, Nuclear Futures Laboratory and Program on Science and Global Security, Princeton University

Walter Robinson, professor, marine, earth, and atmospheric sciences, North Carolina State University

Richard Sandor, CEO, Environmental Financial Products LLC, Chicago, and founder, CCX (the former Chicago Climate Exchange)

Surya Sethi, former principal energy adviser for power and energy, government of India, and former Core Climate Negotiator

Robert Socolow, professor emeritus of mechanical and aerospace engineering, Princeton University

Robert Stavins, professor of business and government, Belfer Center for Science and International Affairs, Harvard Kennedy School

Alf Wills, deputy director-general of environmental affairs and chief climate negotiator, South Africa

I am grateful to my daughter Anna Robinson-Sweet, who helped proofread the book and did the index. Finally, at Yale University Press, I want to thank Joseph Calamia for his expert support and guidance of this project from first to last and Laura Jones Dooley for her painstaking and thorough editorial work.

Index

Abbott, Tony, 106–7
adaptation assistance, 26–28,
    150–52, 167–68, 176–77, 189,
    217n17. *See also* Green Climate
    Fund
Ad Hoc Working Group on the
    Durban Platform (ADP), 163,
    189–90
Agteca-Amazonica, 43
Alcamo, Joseph, 19
Alliance of Small Island States
    (AOSIS), 30, 102–3, 178
Al Sabban, Mohamed, 102
Australia, 51, 106–8
Austria, 137
aviation, emissions from, 17, 75. *See
    also* transportation, emissions
    from

Bali Action Plan, 151
BASIC, 80–81, 92–93, 95–96
Bell, Ruth Greenspan, 55
Benedick, Richard Elliot, 55, 120,
    122, 124, 127–28
Berlin Mandate, 102–3, 108, 129, 188
Bishop, Julie, 107–8

Blechman, Barry, 55
Bloomberg, Michael, 105,
    182
Bodansky, Daniel, 51, 163–68
Boer, Yvo de, 7–8, 148, 154
Boyce, James K., 27
Brazil, 42–43, 93, 95–96, 151–53. *See
    also* BRIC
BRIC (Brazil, Russia, India and
    China), 80–81, 92–93, 95–96. *See
    also individual country names*
British Columbia, 106
Brown, Jerry, 182
Bush, George H. W., 128–29
Bush, George W., 71–72, 109

California, 57–58
Cameron, David, 78–79
Canada, 103–6, 137, 176
cap-and-trade, 46, 50, 74–75. *See
    also* emissions trading
carbon cuts, accomplished, 62, 70,
    75, 135–38
carbon cuts, pledged, 64, 74, 84–88,
    93, 95–96, 104, 106–7, 215n17,
    217n16. *See also* INDC

carbon cutting policies, 19–24; cap-
  and-trade, 46, 50, 74–75;
  common but differentiated
  responsibility (CBDR), 33–36,
  126–27, 138, 142, 147, 150, 168,
  210n7; contraction and
  convergence, 30; free-riding, 56,
  70; hot air, 46, 94, 108–9, 132, 188;
  land use change accounting,
  40–44, 96, 130; mandatory cuts,
  24–25, 73, 109, 126–27, 159;
  offsets, 47–48, 132, 188; pledge
  and review, 24–26, 72–74, 103–4,
  153–54, 164–66, 176, 181; taxation,
  50–52, 56, 106–7, 134, 180, 208n19
carbon tax, 50–52, 56, 106–7, 134,
  180, 208n19
Carlarne, Cinnamon, 39, 67–68, 158
Center for Climate and Energy
  Solutions, 175
Center for Policy Research, 205n29
Chen, Y.-H. Henry, 163–65
Chicago Climate Exchange (CCX),
  46
China, 81–88, 148–52, 157–60, 165,
  183, 204n14. See also BRIC
Clean Air Act, 149, 214n8
Clean Development Mechanism,
  47–49, 132, 180, 188–89
Climate Action Network (CAN),
  108–11, 172, 174, 176. See also
  NGOs
climate catastrophe scenarios, x–xi,
  4, 143, 200n9
climate ethics, 28–33
climate justice, 28–33, 157, 178. See
  also reparations
climate models, 4, 119, 191n5,
  193n22, 200n9
Clinton, Bill, 71–72, 129

Clinton, Hillary, 150, 157
club diplomacy, 55–56
Coalition for Rainforest Nations,
  41
common but differentiated
  responsibility (CBDR), 33–36,
  126–27, 138, 142, 147, 150,
  168, 210n7
Conference of Parties (COP), 35,
  127; COP 1 (Berlin), 72, 102–3,
  108, 129, 183, 188; COP 3 (Kyoto)
  (see Kyoto Protocol); COP 13
  (Bali), 41, 72–73, 151, 181, 189;
  COP 15 (Copenhagen) (see
  Copenhagen climate
  conference); COP 16 (Cancún),
  53, 89, 96, 109, 162, 189; COP
  17 (Durban), 96, 162–63, 189,
  216n2; COP 19 (Warsaw), 163,
  190; COP 20 (Lima), 35, 86, 163,
  168, 179; COP 21 (Paris) (see
  Paris climate conference)
contraction and convergence, 30
Cooper, Richard N., 134
Copenhagen Accord, 150–51, 153–55
Copenhagen climate conference:
  developing countries and, 27–28,
  34, 84, 89, 91, 147; European
  Union and, 158–59; evaluation
  of, 109, 155–62, 189; expectations
  of, 144–45; NGOs and, 146–48,
  154; pledge and review and,
  24–26, 73, 153–54; 2° threshold
  and, 3–4, 151; United States and,
  24, 72–74, 144–45, 148–52, 155–60

Danish draft text, 34, 147, 150,
  214n3
Daschle, Thomas A., 38
Dasgupta, Chandrashekar, 89

Depledge, Joanna, 38–39, 58, 102
developing countries: carbon cutting and, 19–24, 47–48, 81, 130–31, 133; Kyoto Protocol and, 24, 130, 132–33; Paris climate conference and, 176–79, 183; Rio Earth Summit and, 33–36, 81, 100–102, 126–27. *See also* adaptation assistance; Alliance of Small Island States; BRIC; common but differentiated responsibility; G-77; *and individual country names*
diplomacy, other examples of, 53–54, 112–13, 119–25, 138–42
diplomatic personalities, 111–15, 122, 131, 145, 171–72, 182–84
Diringer, Elliot, 163–67

economics of climate diplomacy, 21–25, 194n6, 194n8
emissions trading, 44–50, 67, 74–75, 106, 130, 132, 180–81, 188. *See also* cap-and-trade
Emissions Trading System (ETS), 44, 47, 67
Environmental Defense Fund, 108, 110, 130
Esty, Daniel C., 38
European Union, 61–70, 77–79, 120, 131, 135–37, 158–59, 167, 217n16. *See also individual country names*

Fabius, Laurent, 171, 182, 184
fossil fuels: coal, 11, 63–64, 81–83, 90–91, 93, 106, 193n21; industry lobbying, 8, 102, 110, 125, 180; natural gas, 15, 75, 83, 137, 193n21; oil, 15–16, 101–2, 104–5;

subsidization of, 49, 52–53, 57, 199n58. *See also* carbon cutting policies; greenhouse gases
France, 78, 136, 153, 183–84
Francis (pope), 31, 113–15, 194n7
free-riding, 56, 70
Friends of the Earth, 9, 65, 130, 158, 174

G-77 (Group of 77), 98–102, 167, 177–78, 183
G8 (Group of 8), 4
Gaoli, Zhang, 84–85
Gardiner, Stephen M., 28–29
Gates, Bill, 182
General Agreement on Tariffs and Trade (GATT), 53–54
Germany, 4, 62–65, 78, 136–37, 183
Goldemberg, José, 100
Gore, Al, 71, 109, 129, 131, 189
Green Climate Fund, 28, 168, 189, 214n11, 217n17
greenhouse gases, 15–17, 40, 43–44, 83, 121, 193n21. *See also* aviation, emissions from; fossil fuels; transportation, emissions from
Greenpeace International, 64–65, 109, 130, 173–74
Gupta, Joyeeta, 35–36, 39, 91, 100–101, 129

Hadley Centre for Climate Prediction and Research, 4, 23
Hansen, James, 16, 175, 192n7
Hart, Selwyn C., 20, 27, 35, 168
Heal, Geoffrey, 53
Hedegaard, Connie, 148–49
Hoffman, Matthew J., 39

Hollande, François, 79, 171
hot air, 46, 94, 108–9, 132, 188
human rights, 30–31. *See also*
    climate ethics

ice sheet, melting, ix–x, 4, 174
IEA. *See* International Energy
    Agency
INC. *See* Intergovernmental
    Negotiating Committee
INDC (Intended Nationally
    Determined Contribution), 163,
    165–66, 179, 181–82, 190
India, 81, 88–94, 96–97, 160,
    205n29. *See also* BRIC
Indonesia, 41–42, 81, 96
Intended Nationally Determined
    Contribution. *See* INDC
Intergovernmental Negotiating
    Committee (INC), 123, 128,
    187
Intergovernmental Panel on
    Climate Change (IPCC), 108,
    112, 123, 187
International Energy Agency
    (IEA), 11–12, 52
International Telecommunications
    Union, 54
IPCC. *See* Intergovernmental Panel
    on Climate Change
Iran, 52, 81
Islamic Declaration on Global
    Climate Change, 114
Italy, 136

Jacoby, Henry D., 163–65, 167
Japan, 24, 103–4, 137, 166
Javadekar, Prakesh, 89
Joint Implementation, 132. *See also*
    offsets

JUSCANZ (Japan, United States,
    Canada, Australia and New
    Zealand), 103, 107, 120. *See also*
    individual country names

Kennel, Charles F., 7, 18
Kerry, John, 85, 148–49, 154, 182
Keystone Pipeline, 104–5
Killeen, Timothy, 43
Ki-Moon, Ban, 172
Kinley, Richard, 3, 112, 162
Klaus, Vaclav, 68
Klein, Naomi, 22–23, 25, 38, 67,
    110, 159
Kyoto Protocol: compliance, 35–38,
    211n13; developing countries
    and, 24, 101, 130, 132–33; Eastern
    Bloc states and, 94, 132–33, 137,
    188, 211n14; emissions trading
    and, 44–45; European Union
    and, 62, 69, 135–37; evaluation
    of, 37–40, 133–38, 188, 211n13;
    NGOs and, 130–31; United States
    and, 24, 56–57, 68–74, 109,
    129–33, 138

land use change accounting,
    40–44, 96, 130. *See also* REDD
Lashof, Dan, 108
Laudato Sí, 113–14, 194n7
Leone, Gaetano, 112
Ling, Chee Yoke, 175–76
Lula da Silva, Luiz Inácìo, 145, 151–53

Marrakech Accords, 189
McCarthy, Gina, 182
McKibben, Bill, 6–7, 146–47. *See*
    also 350.org
measuring, reporting, and
    verification (MRV), 178–79

Meese, Ed, 122
Merkel, Angela, 4, 64, 78–79, 115,
    149, 174, 183, 202n36
Mexico, 96
Meyer, Alden, 108–9, 159
mitigation. *See* carbon cutting
    policies
Modi, Narendra, 88–89, 92–94,
    206n33
Mohan, Rakesh, 91
Monbiot, George, 146
Montreal Protocol, 120–22, 187. *See
    also* ozone negotiations, Vienna
    Convention
MRV. *See* measuring, reporting,
    and verification
Mxakato-Diseko, Nozipho Joyce,
    177–78

Naidoo, Kumi, 173
Natural Resources Defense
    Council, 108, 130
Nature Conservancy, 110
Netherlands, 30–31
New Zealand, 103, 165
NGOs, 6–7, 101, 108–11, 130–31,
    146–48, 154, 172
Nordhaus, William, 20–21, 37,
    55–56, 133–34
Norway, 42, 103, 107, 120
nuclear energy, 12, 15, 64, 75, 83, 90,
    200n5
Nuclear Non-Proliferation Treaty,
    138–42

Obama, Barack, 71–78, 115,
    144–45, 150, 152–53, 155–57, 178,
    213n1
offsets, 47–48, 188, 132. *See also*
    Clean Development Mechanism

OPEC (Organization of the
    Petroleum Exporting
    Countries), 51–52, 94, 99, 101–2
Oreskes, Naomi, 27
Orr, Robert, 84–85, 112
Oxfam, 174
ozone negotiations, 67–68, 119–25,
    128, 187

Paris climate conference:
    developing countries and,
    176–79, 183; European Union
    and, 167; evaluation of, 180–81;
    lead-up to, 19, 35–36, 77–79, 107,
    163–69, 171–72; NGOs and, 172;
    2° threshold and, 8, 164–65,
    173–75, 179; United States and,
    35, 77–78, 166–68, 175, 178, 181–82
Pica, Erich, 158
pledge and review, 24–26, 72–74,
    103–4, 153–54, 164–66, 176, 181.
    See also INDC
Podesta, John, 85
Posner, Eric, 29–32
Potsdam Institute for Climate
    Impact Research, 4, 174
protests, xii–xiii, 82, 109–10, 147.
    See also public opinion
public opinion, xi–xii, 43, 65–66,
    76–77, 146, 185–86
Purvis, Nigel, 79

Rahmstorf, Stefan, 4, 21
Ramanathan, V., 16, 179
Rawls, John, 32–33
Reagan, Ronald, 120, 122
REDD (Reducing Emissions from
    Deforestation and Forest
    Degradation), 41–43. *See also*
    land use change accounting

Reggie (Regional Greenhouse Gas Initiative, or RGGI), 44, 47
renewable energy, 8–12, 14–15, 63–65, 83, 86–87, 90–91, 200n5. *See also* nuclear energy; solar energy; wind energy
reparations, 150, 157, 177. *See also* climate justice
Rhineland Model, 67
Rio Earth Summit: accomplishments of, 33–35, 40, 62, 127–29, 187–88; developing countries and, 33–36, 81, 92, 100–102, 126–27; European Union and, 61–62, 126; evaluation of, 37–38, 58; goals of, 3, 37, 126; NGOs and, 108; preparations for, 123–24; United States and, 101, 103, 126, 128–29. *See also* United Nations Framework Convention on Climate Change
Ripert, Jean, 124, 128, 184
Rising Tide Network, 110
Romm, Joseph, 18–19
Rudd, Kevin, 51
Russia, 94–95, 132–33, 206n34, 211n14. *See also* BRIC

Sachs, Jeffrey, 13–14
Sandor, Richard, 44–46
Saran, Shyam, 89–90, 159–60
Sarkozy, Nicolas, 153, 158
Saudi Arabia, 101–2, 183
Scandinavia, 8, 62, 136
Schellnhuber, Hans Joachim, 21, 174
Schunz, Simon, 69
Scripps Institution of Oceanography, 7, 16, 179

Sethi, Surya, 75–76, 160
Shiller, Robert J., 38
Singer, Peter, 29–31, 195n19
solar energy, 10–11, 15, 83
South Africa, 96
Spain, 62–63, 136–37
Stavins, Robert N., 48, 73–74, 181
Stern, Todd, 77, 150, 182
Strong, Maurice, 122–23
Sunstein, Cass, 25, 194n11
Sweden, 50, 133, 136
Switzerland, 103, 107

Thatcher, Margaret, 23, 63, 66–67
350.org, 6–7, 146–47, 175
trade negotiations, 53–54
transportation, emissions from, 11, 15, 91, 204n15. *See also* aviation, emissions from
Trudeau, Justin, 176
Tutu, Desmond, 145–47
2° threshold: diplomatic efforts to adopt, 3–4, 151, 173–75, 189; feasibility of, 6–8, 11–18, 21, 85, 142–43, 164–65, 179; reasoning behind, 4–5, 191n5

UNEP. *See* United Nations Environment Programme
UNFCCC. *See* United Nations Climate Secretariat, Bonn; United Nations Framework Convention on Climate Change
Union of Concerned Scientists, 108, 159, 174
United Kingdom, 62–65, 136–37
United Nations Climate Secretariat, Bonn (UNFCCC), 40–41, 112, 127

United Nations Climate Summit,
New York, xi, 78, 84, 88–89, 185
United Nations Environment
Programme (UNEP), 12, 19, 123,
143, 187
United Nations Framework
Convention on Climate Change
(UNFCCC), 3, 24, 37–38, 62, 81,
126–29, 138–39, 187–88. *See also*
Rio Earth Summit; United
Nations negotiating process
United Nations negotiating
process, 30, 38–40, 50, 57, 111–12
United States: Canada and, 104–5;
China and, 81–82, 84–86, 88;
Copenhagen climate conference
and, 24, 72–74, 144–45, 148–52,
155–60; European Union and, 61,
65, 67–69, 79, 131; Kyoto Protocol
and, 24, 56–57, 68–74, 109,
129–33, 138; Paris climate
conference and, 35, 77–78,
166–68, 175, 178, 181–82; politics

of, 67–68, 70–73, 76–77, 148–49,
154–56; Rio Earth Summit and,
101, 103, 126, 128–29

Victor, David G., 7, 15, 18, 38, 69–70
Vienna Convention, 120–21, 123,
128, 187. *See also* Montreal
Protocol; ozone negotiations

Warsaw Mechanism, 190
Watson, Harlan L., 72–73
Weisbach, David, 29–32
Wills, Alf, 156
wind energy, 8–10, 15, 64–65, 83, 88
Wirth, Timothy E., 38
Wohlstetter, Albert, 140
Wong, Penny, 107
World Bank, 12, 19, 42
World Meteorological
Organization (WMO), 123, 187
World Trade Organization (WTO),
53–54
World Wildlife Fund, 174